STP 1243

Science and Technology of Building Seals, Sealants, Glazing, and Waterproofing: Fourth Volume

David H. Nicastro, Editor

ASTM Publication Code Number (PCN):
04-012430-10

ASTM
1916 Race Street
Philadelphia, PA 19103
Printed in the U.S.A.

ISBN: 0-8031-2008-7
PCN: 04-012430-10

Peer Review Policy

Each paper published in this volume was evaluated by three peer reviewers. The authors addressed all of the reviewers' comments to the satisfaction of both the technical editor(s) and the ASTM Committee on Publications.

To make technical information available as quickly as possible, the peer-reviewed papers in this publication were prepared "camera-ready" as submitted by the authors.

The quality of the papers in this publication reflects not only the obvious efforts of the authors and the technical editors, but also the work of these peer reviewers. The ASTM Committee on Publications acknowledges with appreciation their dedication and contribution to time and effort on behalf of ASTM.

Printed in Scranton, PA
August 1995

Foreword

This publication, Science and Technology of Building Seals, Sealants, Glazing and Waterproofing: Fourth Volume contains papers presented at the symposium of the same name held in Fort Lauderdale, FL on 2–3 February 1994. The symposium was sponsored by ASTM Committee C-24 on Building Seals and Sealants. David H. Nicastro presided as chair is the editor of the resulting publication.

Contents

Overview

The papers published in this book were presented during the Fourth Symposium on the Science and Technology of Building Seals, Sealants, Glazing, and Waterproofing held 2–3 February, 1994 in Fort Lauderdale, FL. The symposium is held annually in conjunction with the winter meeting of ASTM Committee C-24 on Building Seals and Sealants.

This continuing series of symposia provides the best scientific forum in this industry for peer-reviewed research papers. As with most scientific disciplines, significant advances are made by baby steps, and not often by giant leaps. Many of the papers reflect progress reports on on-going research. The open, non-proprietary dialogue between the attendees in instrumental in the continuing betterment of the sealant industry. The questions from the audience during the presentations and the peer-review process of the submitted manuscripts provide some of the input for future adjustments to the direction of the research being conducted.

The eleven papers in this volume represent a good balance of the various membership classes required to participate on ASTM committees: four were contributed by producers, one by a user, and six by general interest members.

The first four papers relate to the durability of sealants, one of the key parameters for selecting a sealant product. Gorman's paper presents data collected from actual field experience by a waterproofing contractor. This rare practical and empirical data can be compared to the various theoretical and laboratory durability evaluation procedures discussed in the next three papers.

Although not correlated with field data, Lacasse and Paroli present interesting and insightful comparisons of the accelerated weathering of five silicone sealants. In the next paper, Lacasse, Margeson, and Bryce evaluate cyclic deformation testing in the laboratory as a means of estimating the fatigue life of another five silicone sealants. This work will be continued on other types of sealant, and the results will be published in a subsequent volume.

The final durability paper, by Beech and Beasley, is an update on their continuing work comparing field weathering and artificial (accelerated) weathering. Their data have been instrumental in "calibrating" the industry's understanding of the acceleration factor that standard weathering tests represent.

Next are presented five papers related to sealant products and applications. The paper by O'Neil, Klosowski, Altes, and Bergman is first not only in order of publication, but represents the first research addressed solely at sealant aesthetic issues. In addition to the excellent experimental program discussed, the paper is also a useful resource because it presents a nearly comprehensive discussion on sealant aesthetic issues. The only common aesthetic problems not covered by their paper are sealant color change and substrate water staining near joints (which often resembles one of the four mechanisms discussed in their paper). More importantly, their paper should cause specifiers great concern: some specimens developed latent staining which were not predicted by the standardized tests currently employed.

Myers' paper on the performance of "Band-Aid" (or Bridge Seal) joints provides qualitative and quantitative recommendations for designing seals in this configuration. Furthermore, the stresses developed in "band-aid" joints are compared to his previous findings on butt and fillet joints.

Although Abo-Qaidis and Al-Qadi were unable to present their paper at the symposium, we are fortunate to have their published manuscript. Their elegant test program authoritatively indicates the relative performance of several joint types in reducing corrosion in reinforced concrete pavements.

Awareness of the need to protect the environment continuously increases. Lee's paper is therefore a landmark because it is the first to specifically address the use of a construction sealant to benefit the environment.

Polysulfide-based sealants are generally recognized as having been the first type to be classified as "high-performance"—that is, high movement capability and long durability in exterior construction joints. Fiorillo and Harding present new developments in polysulfide technology, which indicate that these formulations can be effective even without primer. Although not discussed in their paper, priming is still a valuable step in promoting the durability of a sealant joint. Nevertheless, good unprimed performance is essential because many projects still do not employ primer, even when it is specified.

The last two papers relate to glazing. Wilson discusses a case history of reglazing a building with defective structural sealant design and installation. Considering that sealant adheres the glass lites to the frames in four-sided structural glazing, this type of failure represents a real hazard. Wilson's methodology can be applied to other projects to evaluate their existing condition and remedy them if necessary.

Finally, Lowe's paper on Insulated Glass Unit (IGU) testing in Europe is mandatory reading for anyone planning to sell IGUs in the world market after 1998. As international standards gain acceptance, those who are involved in their development will have a distinct comprehensive advantage.

Even as we publish this volume, I look forward to the next series of papes in this dynamic industry. I encourage all readers to participate in the standards development process, to attend the symposia, and to contribute new papers. By your participation and feedback, the industry is advanced and we all benefit from improvements to our constructed environment.

David H. Nicastro
Nicastro Enterprises
Houston, TX; symposium
chair and editor.

Durability

Patrick D. Gorman[1]

WEATHERING OF VARIOUS SEALANTS IN THE FIELD - A COMPARISON

REFERENCE: Gorman, P. D., **"WEATHERING OF VARIOUS SEALANTS IN THE FIELD - A COMPARISON,"** Science and Technology of Building Seals, Sealants, Glazing, and Waterproofing - Fourth Volume, ASTM STP 1243, David H. Nicastro, Ed., American Society for Testing and Materials, Philadelphia, 1995.

ABSTRACT: Many controlled environment tests have been performed trying to predict weathering of sealants in the field. This paper will look at over 220 samples of sealants cut from field sites. The sealants range in age from one to 33 years. The samples are compared for weathering characteristics of sealants on different building elevations, painted versus unpainted sealants, polysulfides versus polyurethanes or silicones, and a variety of joint sizes. Field adhesion tests were done on each sample. Durometer (Shore A) hardness data shows relative hardness of the samples.

Most samples were taken from the desert climate of El Paso, Texas. High ultra violet radiation and heat in this area give insight to compare against earlier laboratory studies. Lack of ambient or surface moisture on samples that exhibited "reversion" suggests that this phenomenon can occur without moisture.

The information collected here will provide field data for future studies on the subject of sealant weathering. It also offers a valuable data base for evaluating other publications on the subject.

KEYWORDS: Polyurethane, polysulfide, silicone, reversion, durometer, sealant alligatoring, sealant tackiness, sealant color, sealant age, profile.

[1]Mr. Gorman is President of Gorman Moisture Protection, Inc., 2115 E. Missouri, El Paso, Texas 79923.

INTRODUCTION

Much work has been done to simulate the environment in which a joint sealant must survive. The simulations are intended to provide a controlled reproduction of the problems perceived in the field. Establishing a baseline of field weathering characteristics is essential to addressing the solutions. Therefore, several samples were pulled from buildings and their data recorded.

Method

Sealant samples were pulled from two sides of each building. Wherever possible, sealants were taken from the north and south sides.

Most sites were selected from completed projects of the author's application company from which original file information was available. Some samples were recorded from projects that did not have historical information available. In these cases, research was done to estimate the date of application and generic type of material. An effort was made to select sites from a list of completed projects without any regard to known conditions. In one case, however, two reverted samples were taken from a project sealed by a different company.

The following information was recorded on each of the 223 samples:

- Date of application
- City from which the sample was taken
- Photographs taken
- Backup material (bond breaker tape, closed cell, open cell, other)
- Distance between joints
- Substrate type
- Sealant original color
- Sealant type (acrylic, polysulfide, polyurethane, or silicone)
- Sample profile measurements:
    - Thickness in the middle
    - Thickness at the bond line
    - Joint width
- Orientation (north, south, east, west, shaded, or painted)
- Appearance (alligatored, chalky, cracks, dirty, ok, or reverted)
- Tackiness on a scale of 1 to 5 (1=sticky, 5=dry)
- Durometer hardness (Shore A)
- Distance pulled to adhesive or cohesive failure

Most samples are from the desert climate of El Paso, Texas, where the average ambient humidity is 28 percent. Extreme temperatures range from 44.4^{0}C to -22.2^{0}C (112^{0}F to -8^{0}F). Average rainfall is only 19.8 cm (7.82 inches) per year. Buildings in this area get sun shine 293 days a year. Forty degree daily ambient temperature swings are not uncommon here.

OBSERVATIONS

Age of Samples

There were two samples pulled from a building sealed in 1961. Most were from four to eighteen years old (Figure 1.).

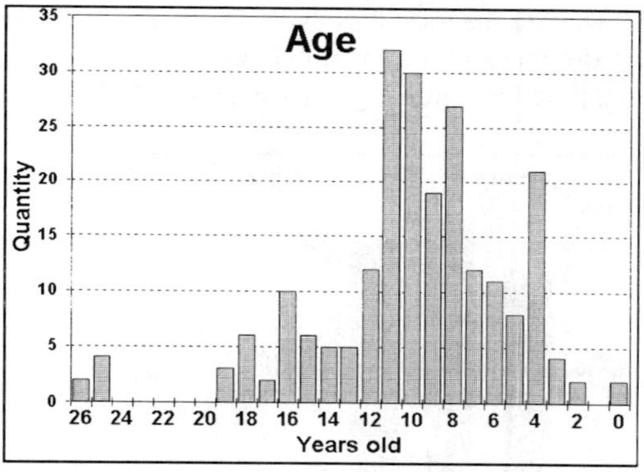

Figure 1

Backup Materials

There were five categories of backup materials. The "other" usually consisted of asphalt impregnated fiber board. Most of samples had open cell rod. The breakdown is shown in Figure 2.

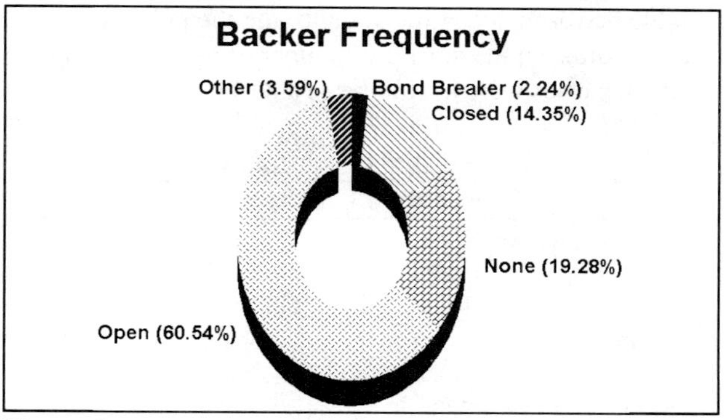

Figure 2

There was no evidence of weathering effects due to the type of backup materials. Most samples of reverted sealant had dry backup materials at the time of removal. There is no question that the type and placement of backup materials impacted performance of the sealants pulled. In one sample, twisted closed cell rod had begun to unwind as the joint cycled but before the single component polyurethane cured.

Joint Spacing

The distance between adjacent joints to the samples may be significant when evaluating stress cracks. A preponderance of samples occurred in joints spaced between three to nine meters (10 to 30 feet) (Figure 3.). Originally the field sample worksheet included a space to record whether the joint was dynamic or static. This became a question of degree and was dropped since all would have some movement. One sample of a silicone was taken from a parking garage with a spacing of 41 meters (135 feet).

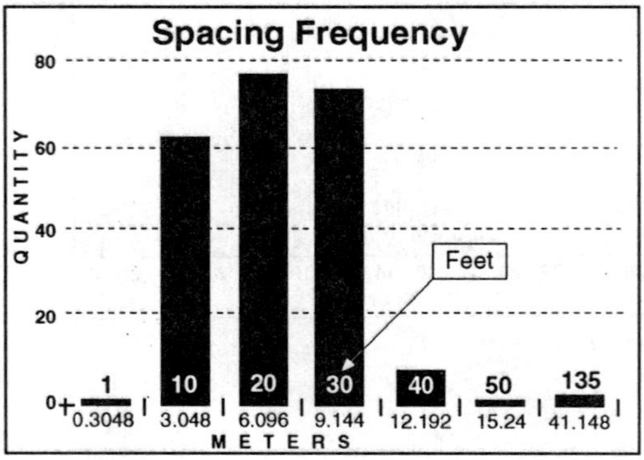

Figure 3

Substrates

Four basic substrates were recorded. There was evidence of weathering relative to substrates though probably more due to color. The type and soundness of the substrates had a direct bearing on the performance of the sealants and the pull test (discussed later). According to the project records, an insignificant number of the samples had been applied over a primer. The four categories were as follows (Table 1):

TABLE 1--Types of substrates

Substrates	Quantity	Percent
1. Brick	45	20.2%
2. Concrete	152	68.2%
3. Metal to Masonry	17	7.6%
4. EIFS	9	4.0%
Total	223	100.0%

Sealant Colors

Gray was the most frequent color of the samples. Tan, white, redwood, black, and bronze are all represented as follows (Table 2):

TABLE 2--Sealant colors tested

Color		Quantity	Percent
1. Black		6	2.7%
2. Bronze		10	4.5%
3. Gray		105	47.1%
4. Redwood		19	8.5%
5. Tan		53	23.8%
6. White		30	13.5%
	Total	223	100.0%

Sealant Types

TABLE 3--Types of sealants pulled

Sealant Type		Quantity	Percent
Acrylics		3	1.3%
Polysulfides		21	9.4%
Polyurethanes		185	83.0%
Silicones		14	6.3%
	Total	223	100.0%

TABLE 4--The brands (coded) of sealants pulled

Sealant Type	Brand code	Qty	Percent of category	Percent of total
Acrylics:	1	3	100%	1.3%
Polysulfides:	10	11	52.4%	4.9%
	13	8	38.1%	3.6%
	24*	2	9.5%	0.9%
Polyurethanes:	2	103	55.7%	46.2%
	7	2	1.1%	0.9%
	8	30	16.2%	13.5%
	9	3	1.6%	1.3%
	14	12	6.5%	5.4%
	15	1	0.5%	0.4%
	17	3	1.6%	1.3%
	20	4	2.2%	1.8%
	22*	27	14.6%	12.1%

TABLE 4--The brands (coded) of sealants pulled (Continued)

Sealant Type	Brand code	Qty	Percent of category	Percent of total
Silicones:	3	8	57.1%	3.6%
	4	4	28.6%	1.8%
	23*	2	14.3%	0.9%
Total	223			100.0%

*Brands 22, 23 and 24 are unknown beyond generic type and may represent multiple products.

Acrylics occurred on only one building. The sample was not large enough to be useful. Polysulfides and silicones were fairly well represented. Polyurethanes made up a high majority of the samples and the data derived is therefore most useful. One brand of polyurethane represented fifty-six percent of all polyurethanes and forty-six percent of all samples. Even though two brands of urethanes made up seventy two percent of all urethanes, their ages did not overlap, making comparison insignificant (Figure 4).

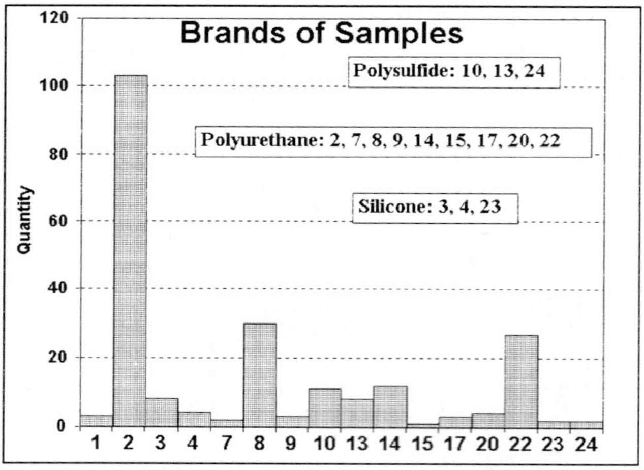

Figure 4--Brands 22, 23 and 24 are unknown beyond generic type and may represent multiple products.

Sample Profile Measurements

Joint dimensions bear on weathering of sealants. On two occasions, an attempt was made to pull samples of a polyurethane from false joints in precast concrete. They were alligatored to the full depth and heavily chalked. Heat may have contributed to this complete deterioration.

Three measurements were taken of each sample. Thickness refers to the measurement in the middle of the sample section. Most samples ranged between .318 cm(.125 in.) and 1.27cm (.5 in.) thick.

Width is the measurement between the two bond lines. The dominant range of widths was between 1.27 cm (.5 in.) and 1.91 cm (.75 in.).

Thickness at the sides (often referred to as "bite") means the depth of the bond line of the sealant to the substrates. Mostly, the range for this measurement was between 1.27 cm (5 in.) and 3.18 cm (1.25 in.).

The manufacturers' recommendation, in terms of depth to width for the sealants in this study, is fifty percent. Most of the samples ranged between twenty and eighty percent. The majority was measured to be fifty percent as shown in Figure 5.

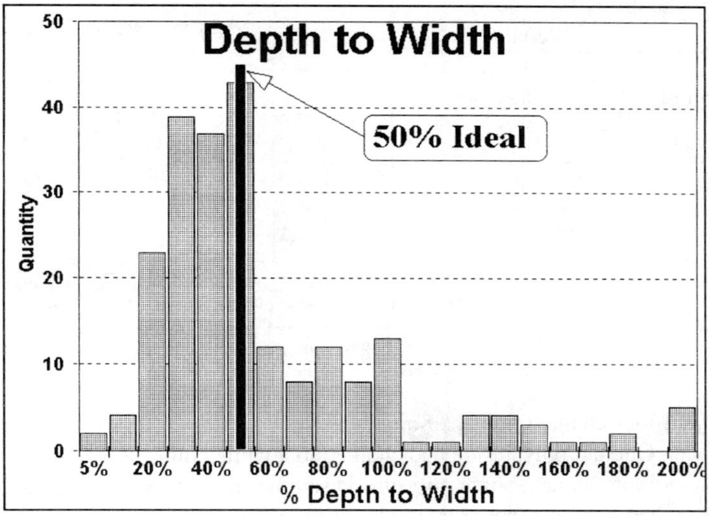

Figure 5

Orientation

The samples were broken into six categories of orientation. In addition to north, south, east, and west are shaded and painted categories. Thirty-one percent of the samples were painted (Table 5).

TABLE 5--Categories of orientation

Orientation	Excluding painted samples:	Orientation	Including painted samples:
North	38%	North	25%
South	38%	South	26%
East	6%	East	4%
West	16%	West	12%
Shaded	2%	Shaded	2%
		Painted	**31%**

Appearance

Appearance was reported as either alligatored, dirty, chalky, cracked, reverted, or ok. There were 14 cases which required two designations (for example, a sample can be both dirty as well as alligatored). Those designated "ok" were in good to excellent condition. These samples had none of the five other characteristics (Table 6 and Figure 6).

TABLE 6--Sampling breakdown

Type	Percent
Ok	41%
Alligatored	28%
Dirty	9%
Reverted	8%
Cracks	8%
Chalky	6%

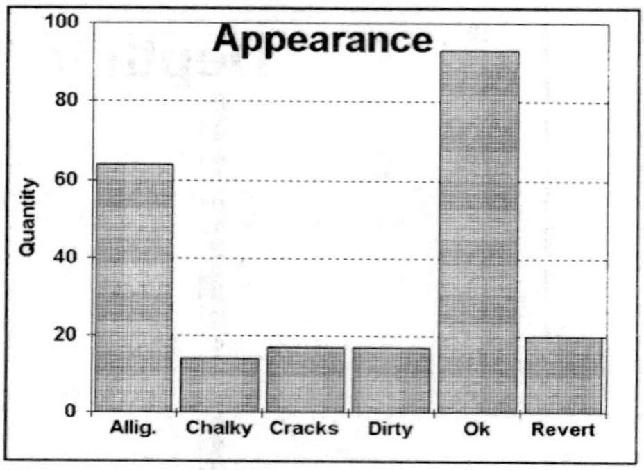

Figure 6

Polysulfides chalked lightly. Some white polysulfides (30 years old) became tan on the surface. One tan polysulfide had chalked to a white color. Polysulfides did not alligator like polyurethanes. Some polysulfide samples had cracked while others on the same structure had not. There was no evidence of excessive structural movement on the cracked polysulfide joints (Photo No. 1).

Polyurethanes showed the greatest change of appearance with age. There was a vast difference in appearance with age among the three types of sealants. The polyurethane samples had alligatored (sometimes very badly), chalked, picked up dirt, or faded. The best aged polyurethanes were those covered by paint or shaded. Most polyurethane data sheets advise against painting sealants since paint would crack during movement. Alligatored sealant often appeared less appealing than cracked paint over the sealant in this study (Photos No. 2 to 5).

It is important to note that if all painted samples are removed from the chart data, the category "Ok" loses 77% of its entries. Without the painted samples the heavy weight shifts to the "Alligatored" category. The breakdown is (Table 7):

TABLE 7--With and without paint

	Allig.	Chalk	Crack	Dirt	Ok	Revert
With paint:	29%	7%	7%	4%	**43%**	10%
Without:	**44%**	11%	10%	6%	15%	15%

Of the 19 reverted samples, 12 came from the south side, 3 each from the north and west, and one from an interior joint (this author defines reversion as the condition in which a cured sealant rubber reverts to a non-elastic paste to some degree). The interior sample was from a tile joint in the floor of a milk processing plant. That was one of only two samples exposed to any moisture. The temperature in that room was always between 27°C (80°F) and 32°C (90°F).

Another sample of a structure with reverted sealant was a single story precast concrete bank with adjacent lease spaces, sealed in December 1982. It had a courtyard and was located in the desert. All sealant that was touched by sunlight was in varying stages of reversion. All sealant that was completely shaded (e.g. soffits in covered walks) was sound rubber. Condensation was not a factor due to the low average relative humidity and mild winters (Figure 7).

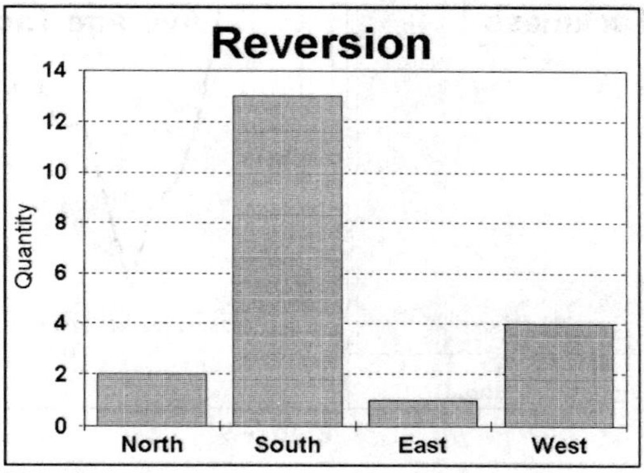

Figure 7

Silicones were reported at 50% dirty and 50% ok. From a visual perspective, the silicones did not seem to change in appearance other than a light dirt pick up on half of the samples. Those samples listed as dirty included some traffic joints (Photo No. 6).

Tackiness

Tackiness of the samples was reported on a scale of one to five. One meant that the sealant would stick to a finger and leave a residue when pulled off. Five meant that the sealant would have no sticky feel at the cut edge whatsoever. Though most received a marking of five, thirty five (16%) were rated one or two. The actual ranking was (Table 8):

TABLE 8--Tackiness scale

Tackiness	Scale	Quantity	Percent
Sticky	1	16	7%
	2	19	9%
Medium	3	25	11%
	4	45	20%
Dry	5	118	53%

There was a slight increase in tackiness with age of samples. All silicones were dry (5). The only correlation of tackiness to the six appearance categories was the reverted samples (Figures 8 and 9).

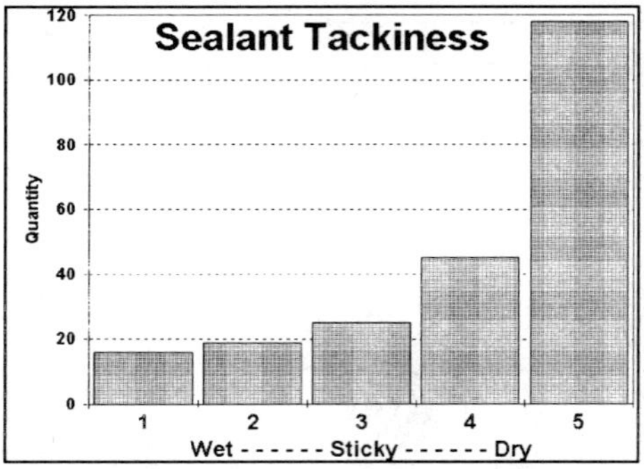

Figure 8

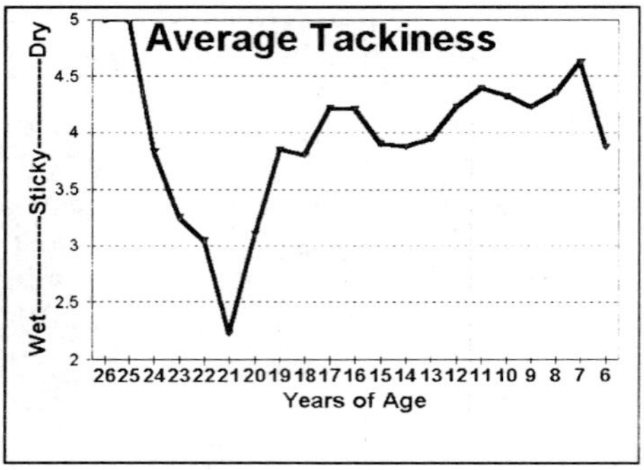

Figure 9

Durometer Hardness (Shore A)

Hardness was established using a Shore A durometer. The samples were measured within two days of being pulled, after achieving room temperature. Samples were cut to about 6.35mm (0.25 inches) square, placed on a scale, manually pressed to 1.36 kilograms (three pounds), and held for five seconds at which time the measurement was taken. Each sample was measured in this manner a minimum of three times.

The majority (80%) of the samples measured in the range between 15 and 35 durometer (Shore A) as seen in Figure 10.

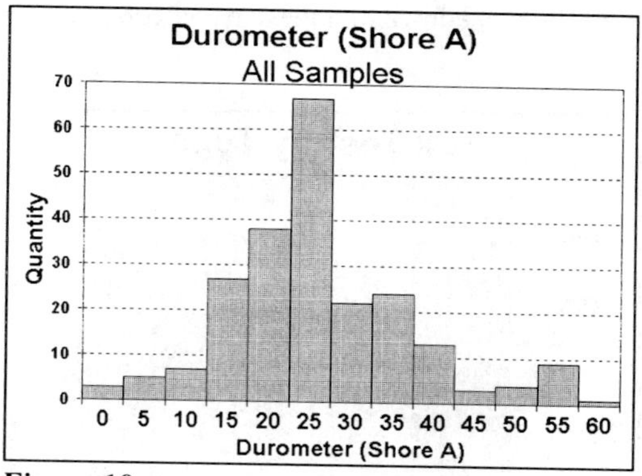

Figure 10

Distance Pulled to Failure

The final aspect recorded is the distance each sample was pulled from the substrate before failure occurred. Failure was reported as either adhesive of cohesive. The procedure was to cut a tab, grasp it one inch from the substrate, and slowly pull at a ninety degree angle until failure occurred. The measurement was made at the point of failure.

The resulting data does not identify why a sample failed adhesively or cohesively. The ability of a sealant to stick is dependent on factors beyond the scope of this study, such as the substrate soundness or application technique.

Those that failed adhesively usually did so due to substrate failure. This was particularly prevalent on exposed aggregate precast concrete in which the sealant was applied close to the panel face. There is some correlation between adhesive failure and Shore A hardness on samples over 40 durometer (Figure 12). Weak substrates caused failure even with softer sealants.

Figure 11 shows the breakdown of adhesive and cohesive failure by orientation. Figure 12 shows the breakdown of adhesive and cohesive failure by hardness.

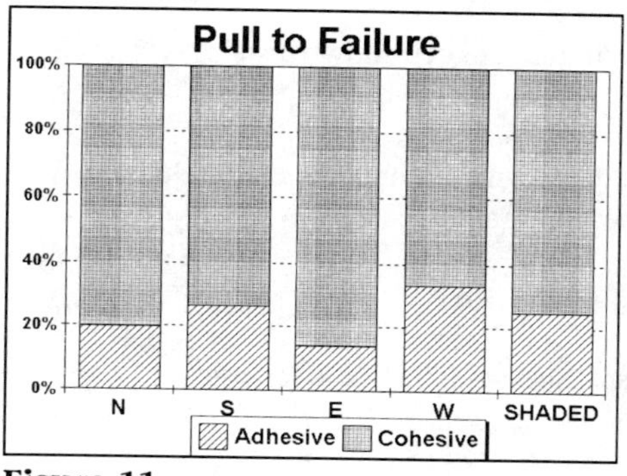

Figure 11

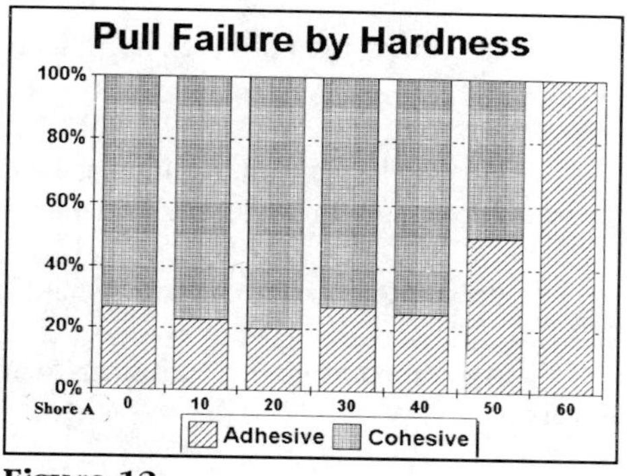

Figure 12

Figure 13 shows the breakdown of adhesive and cohesive failure by generic type of sealant.

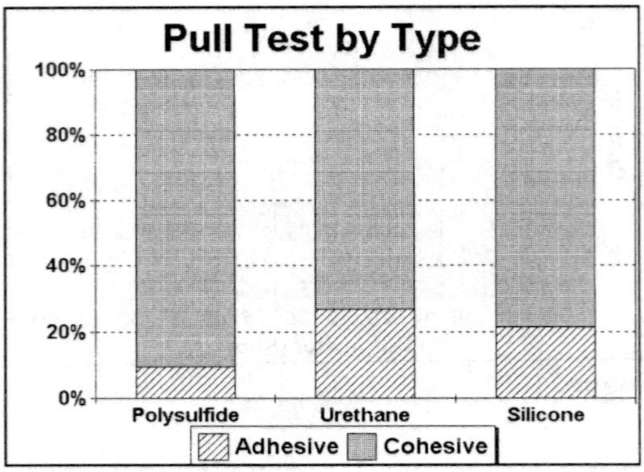

Figure 13

Conclusions

The exposed southern elevation of a building will see the adverse effects of weathering at an accelerated rate in a desert climate.

Painted polyurethanes do not alligator where paint has not broken. While painting an elastomeric sealant is not generally recommended due to the paints' lack of elasticity, it does tend to prolong the service life of the sealant considerably.

Some polysulfides tan while others chalk to a white color in the weathering process.

Some polyurethanes can revert with no other source of moisture than very low ambient humidity. This observation was restricted to one brand in this study.

Other than light dirt pick up, silicones in this research show very little effects of weathering.

Polyurethanes tend to alligator with UV exposure, though the cracks do not tend to be deep enough to hinder performance except in shallow beads.

Polyurethanes become somewhat tackier with age.

Age or exposure does not seem to affect durometer hardness (excluding reversion).

Durometer hardness is not a factor in a field pull test up to 40 (Shore A). Beyond 40, adhesion suffers.

The very hard sealants tended to have adhesive failure.

Durometer hardness versus time indicated the majority of the sealants retained flexibility and did not harden or embrittle.

SUPPORTING DATA

Table 9 is the entire data spreadsheet. The legend is as follows:

Column 1. The date the sealant application was completed.

Column 2. Type of backup material found.
1 is bond breaker tape.
2 is closed cell backer rod.
3 is none found (usually 3-sided adhesion to concrete).
4 is open cell backer rod.
5 is other (e.g. fiber board, plastic from strips).

Column 3. Joint spacing in meters and feet.

Column 4. Type of substrate.
1 is brick
2 is concrete.
3 is exterior insulation and finish system (EIFS).
4 is metal to masonry.

Column 5. Sealant color.
1 is black.
2 is bronze.
3 is gray.
4 is redwood.
5 is tan.
6 is white.

Column 6. Brand of sealant.
1 is acrylic.
10, 13, and 24 are polysulfides.
2, 7, 8, 9, 14, 17, 20, and 22 are polyurethanes.
3, 4, and 23 are silicones.

Column 7. Sealant sample dimension measured in the center of the cross section and reported in centimeters and inches.

Column 8. Sealant sample dimension measured at the bond line at the sides of the joint and reported in centimeters and inches.

Column 9. Sealant sample dimension measured across the width of the joint and reported in centimeters and inches.

Column 10. Orientation from which the sample was taken.
1 is north.
2 is south.
3 is east.
4 is west.
5 is shaded.
6 is painted.

Column 11. Appearance of the samples.
1 is alligator cracks.
2 is chalky.
3 is cracks (different from alligator cracks in that these are parallel as opposed to reticulated cracks as in alligatoring).
4 is dirt pick up.
5 is ok (from good to excellent).
6 is reverted sealants.

Column 12. Same as 11. above. In some cases two designations were given to one sample.

Column 13. Tackiness on a scale of 1 = wet to 5 = dry.

Column 14. Durometer (Shore A) hardness.

Column 15. Field pull-to-failure test reported in centimeters and inches.

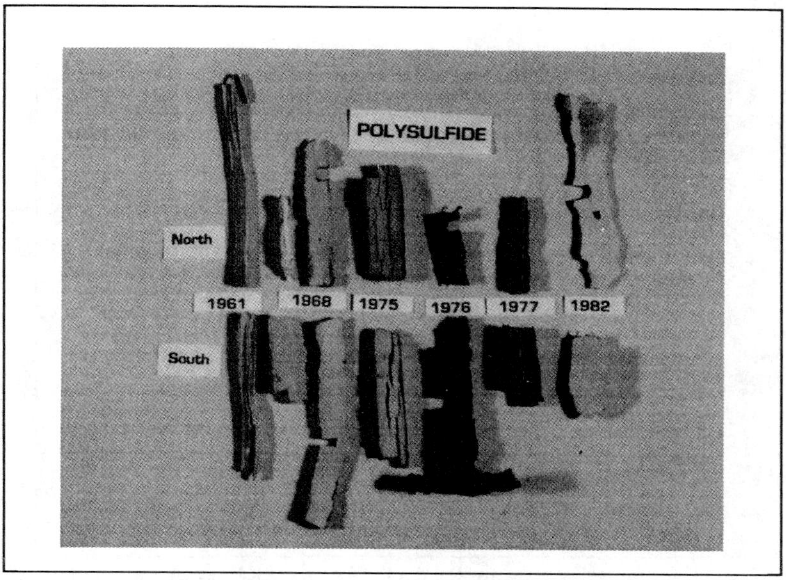

Photo No. 1--Polysulfide examples from 1961 to 1982, north and south exposures.

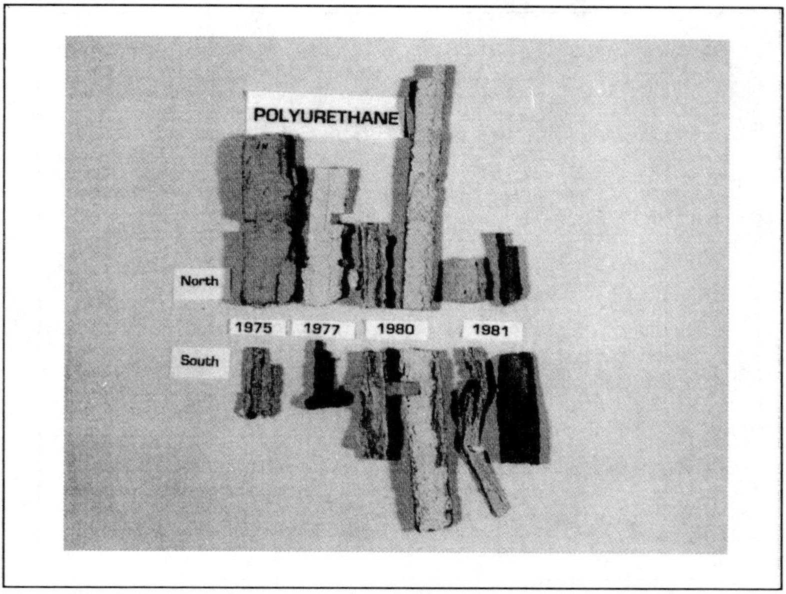

Photo No. 2--Polyurethane examples from 1975 to 1981, north and south exposures.

Photo No. 3--Polyurethane examples from 1982 to 1984, north and south exposures.

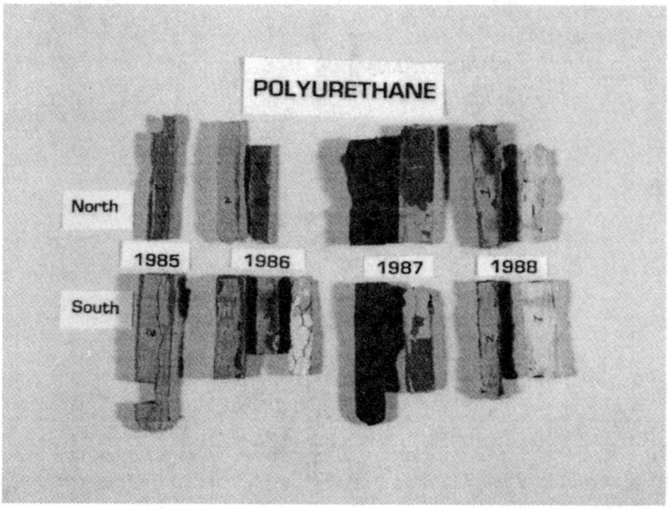

Photo No. 4--Polyurethane examples from 1985 to 1988, north and south exposures.

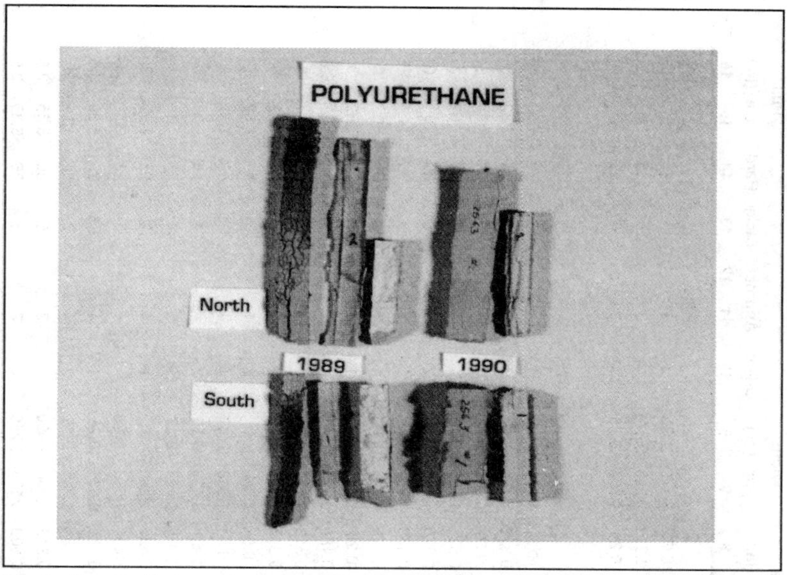

Photo No. 5--Polyurethane samples from 1989 and 1990, north and south exposures.

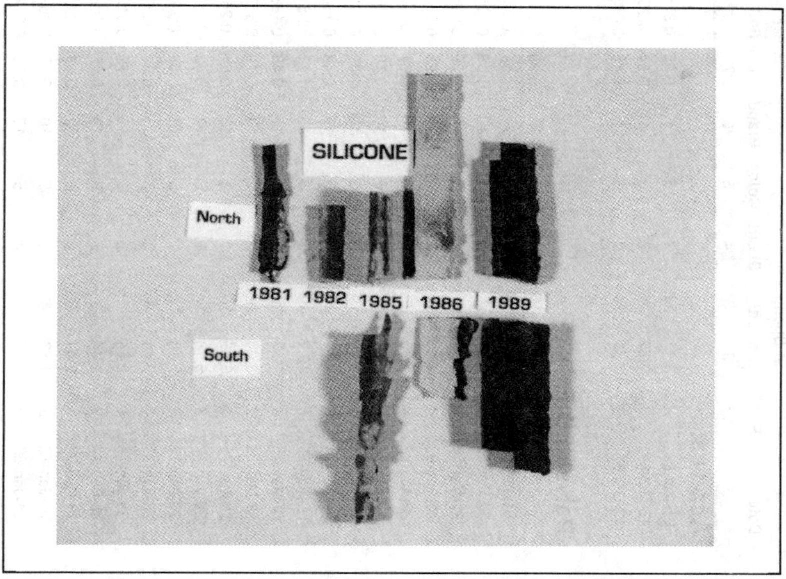

Photo No. 6--Silicone samples from 1981 to 1989, north and south exposures.

TABLE 9--Database

Date	Rod	Jt Spcng m	(ft)	Sbstrt	Color	Brand	Center cm	(in.)	Bite cm	(in.)	Width cm	(in.)	Orient.	Apprnc		Tacky	Hard	Pull cm	(in.)
1	2	3	3	4	5	6	7	7	8	8	9	9	10	11	12	13	14	15	15
Acrylics:																			
01-Jan-82	2	6.71	22	2	5	1	0.95	0.375	1.905	0.75	3.175	1.25	3	3		5	25	0	0
13-Oct-83	2	7.32	24	2	6	1	0.64	0.25	0.795	0.313	2.223	0.875	1	3		5	55	0	0
13-Oct-83	2	7.32	24	2	6	1	0.16	0.063	0.635	0.25	1.905	0.75	2	3		5	55	0	0
Polysulfides:																			
01-Jan-61	5	4.57	15	2	6	24	0.95	0.375	1.27	0.5	1.588	0.625	1	4		5	39	1.27	0.5
01-Jan-61	4	5.49	18	1	6	24	0.64	0.25	0.794	0.313	1.27	0.5	3	5		5	54	7.62	3
20-Mar-68	3	2.44	8	2	6	10	1.59	0.625	1.905	0.75	0.953	0.375	2	5		5	36	7.62	3
20-Mar-68	4	1.83	6	2	6	10	1.59	0.625	1.588	0.625	1.905	0.75	1	4		5	35	7.62	3
20-Aug-68	4	1.83	6	2	6	10	0.95	0.375	0.953	0.375	1.905	0.75	2	5		5	35	0	0
20-Aug-68	4	1.83	6	2	6	10	0.64	0.25	1.111	0.438	1.905	0.75	1	5		5	35	11.4	4.5
20-Jun-74	4	1.52	5	4	6	10	0.79	0.313	0.953	0.375	0.953	0.375	4	1		2	6	0	0
20-Jun-74	4	5.79	19	4	6	10	0.32	0.125	0.635	0.25	1.27	0.5	1	1		3	14	7.62	3
20-Jun-74	3	7.32	24	1	6	10	0.48	0.188	0.476	0.188	0.953	0.375	1	1		3	4	1.27	0.5
16-May-75	4	12.2	40	2	3	10	1.27	0.5	0.794	0.313	3.175	1.25	1	3		5	26	12.7	5
16-May-75	4	12.2	40	2	3	10	0.48	0.188	0.794	0.313	1.905	0.75	2	3		5	21	15.2	6
08-Mar-76	2	3.05	10	2	2	13	0.79	0.313	1.27	0.5	2.54	1	2	1		3	21	7.62	3
08-Mar-76	2	3.05	10	2	2	13	0.64	0.25	1.27	0.5	1.588	0.625	1	4		1.5	19	7.62	3
20-Jan-77	3	5.18	17	1	6	13	0.48	0.19	1.27	0.50	0.476	0.19	2	4		4	15	0	0
20-Jan-77	3	5.18	17	1	6	13	0.64	0.25	0.476	0.19	0.476	0.19	1	5		4	15	0	0
28-Jan-77	4	2.44	8	2	3	13	0.64	0.25	1.588	0.625	2.54	1	2	4	1	2	23	22.9	9
28-Jan-77	4	2.44	8	2	3	13	0.79	0.313	1.27	0.5	1.588	0.625	1	4	1	3	18	20.3	8
23-Feb-77	3	2.44	8	2	6	13	0.48	0.188	0.476	0.188	0.635	0.25	1	4	1	1.5	25	8.89	3.5
23-Feb-77	4	2.44	8	2	6	13	0.64	0.25	1.27	0.5	1.905	0.75	2	4	1	2	21	10.2	4
01-Jan-82	2	1.83	6	2	5	10	0.16	0.063	0.794	0.313	2.54	1.000	1	2		5	36	6.35	2.5
01-Jan-82	2	1.83	6	2	5	10	0.79	0.313	1.27	0.500	3.175	1.250	2	2		5	36	6.35	2.5

TABLE 9--Database (Continued)

Date	Rod	Jt Spcng m	(ft)	Sbstrt	Color	Brand	Center cm	(in.)	Bite cm	(in.)	Width cm	(in.)	Orient.	Apprnc		Tacky	Hard	Pull cm	(in.)
1	2	3	3	4	5	6	7	7	8	8	9	9	10	11	12	13	14	15	15
Silicones:																			
20-Nov-78	3	7.92	26	2	3	3	0.64	0.25	0.635	0.25	0.318	0.125	1	5		5	23	7.62	3
20-Nov-78	3	7.92	26	2	3	3	0.64	0.25	0.635	0.25	0.318	0.125	3	5		5	21	7.62	3
20-Nov-78	3	7.92	26	2	3	3	0.79	0.313	0.635	0.25	0.635	0.25	4	5		5	27	6.35	2.5
29-May-81	3	1.22	4	4	1	4	0.32	0.125	0.318	0.125	1.111	0.438	1	4		5	47	1.27	0.5
08-Sep-82	4	3.66	12	2	3	3	1.27	0.5	0.476	0.188	1.27	0.5	2	5		5	13	3.81	1.5
01-Jan-85	4	9.75	32	2	3	23	0.95	0.375	1.905	0.75	1.905	0.75	2	4		5	13	2.54	1
01-Jan-85	3	9.75	32	2	3	23	2.54	1	3.175	1.25	1.27	0.5	1	4		5	18	0	0
21-Mar-85	3	1.83	6	4	6	4	0.64	0.25	1.27	0.5	1.905	0.75	2	5		5	34	0	2
13-Jan-86	1	4.88	16	4	3	3	0.64	0.25	0.953	0.375	3.175	1.25	1	4		5	18	7.62	3
13-Jan-86	1	4.88	16	4	3	3	0.48	0.188	0.635	0.25	3.175	1.25	2	4		5	16	7.62	3
03-Feb-86	3	2.44	8	4	1	4	0.79	0.313	1.27	0.5	1.905	0.75	4	4		5	38	3.81	1.5
24-Feb-86	3	1.83	6	4	3	4	0.32	0.125	0.318	0.125	1.27	0.5	1	4		5	48	1.27	0.5
01-Jan-89	2	41.1	135	4	3	3	0.97	0.38	1.905	0.75	4.775	1.88	4	5		5	18	1.27	0.5
01-Jan-89	2	41.1	135	2	3	3	0.74	0.29	1.6	0.63	3.175	1.25	4	5		5	25	0	0

TABLE 9--Database (Continued)

Date	Rod	Jt Spcng m	(ft)	Sbstrt	Color	Brand	Center cm	(in.)	Bite cm	(in.)	Width cm	(in.)	Orient.	Apprnc		Tacky	Hard	Pull cm	(in.)
1	2	3	3	4	5	6	7	7	8	8	9	9	10	11	12	13	14	15	15
Polyurethanes:																			
12-Apr-75	3	4.57	15	1	5	22	0.64	0.25	0.953	0.375	2.54	1	2	5		1	14	12.7	5
12-Apr-75	3	0.3	1	4	3	22	1.27	0.5	1.27	0.5	0.794	0.313	2	5		5	26	0.64	0.25
12-Apr-75	5	2.44	8	1	5	22	0.95	0.375	1.27	0.5	4.288	1.688	1	5		2	19	10.2	4
12-Apr-75	4	4.57	15	2	3	22	0.64	0.25	0.953	0.375	1.27	0.5	2	2		5	38	0	0
30-Jun-77	3	2.44	8	2	3	17	2.54	1	2.54	1	1.27	0.5	1	5		1.5	48	0	0
30-Jun-77	3	2.44	8	2	3	17	2.22	0.875	2.223	0.875	1.27	0.5	2	5		1	55	10.2	4
20-Oct-77	4	11	36	1	5	22	0.64	0.25	0.476	0.188	3.175	1.25	1	5		2	15	0	0
20-Oct-77	5	10.1	33	1	5	22	0.64	0.25	0.635	0.25	1.27	0.5	2	6		1	11	0	0
27-Nov-78	5	6.1	20	2	3	17	0.64	0.25	0.794	0.313	1.429	0.563	2	2		5	38	8.26	3.25
27-Nov-78	3	4.88	16	1	5	22	0.95	0.375	0.953	0.375	0.794	0.313	1	6		2	8	7.62	3
27-Nov-78	4	4.88	16	1	5	22	0.95	0.375	1.27	0.5	1.905	0.75	2	6		2	11	0	0
21-Jun-79	4	7.32	24	2	5	14	0.32	0.125	0.318	0.125	1.905	0.75	1	3		5	26	2.54	1
21-Jun-79	4	7.32	24	2	5	14	0.32	0.125	0.476	0.188	1.27	0.5	2	3		5	40	2.54	1
31-Aug-79	3	6.1	20	2	3	14	0.48	0.188	0.794	0.313	1.27	0.5	1	5		1.5	35	6.35	2.5
31-Aug-79	4	6.1	20	2	3	14	0.48	0.188	0.476	0.188	1.27	0.5	2	5		2	37	0	0
12-Oct-79	3	4.57	15	1	6	14	1.91	0.75	2.223	0.875	1.27	0.5	1	5		5	22	10.2	4
18-Apr-80	4	6.71	22	2	6	14	0.32	0.125	0.476	0.188	1.905	0.75	1	1	4	5	26	7.62	3
18-Apr-80	4	6.71	22	2	6	14	0.32	0.125	1.27	0.5	1.905	0.75	2	1	4	5	25	0	0
19-Sep-80	4	6.1	20	1	3	14	1.11	0.438	1.111	0.438	0.635	0.25	1	3		4.5	52	3.81	1.5
02-Oct-80	4	6.4	21	2	3	14	0.79	0.313	0.635	0.25	2.54	1	1	5		1	35	0	0
02-Oct-80	4	6.4	21	2	5	2	0.64	0.25	0.953	0.375	3.175	1.25	2	5		4	21	0	0
29-May-81	3	4.57	15	2	3	2	0.79	0.313	0.318	0.125	1.905	0.75	1	2	1	4.5	23	1.27	0.5
17-Jul-81	4	2.74	9	2	5	2	0.48	0.188	1.27	0.5	1.905	0.75	1	5		4.5	23	5.08	2
17-Jul-81	4	2.74	9	2	5	2	0.16	0.063	0.318	0.125	1.27	0.5	2	1		3.5	21	0	0
20-Jul-81	4	1.83	6	2	3	14	0.32	0.125	0.476	0.188	1.588	0.625	1	1		4.75	39	6.99	2.75
20-Jul-81	4	2.44	8	1	5	2	0.48	0.188	0.953	0.375	1.746	0.688	2	1		4.5	21	0	0
20-Jul-81	4	11.9	39	2	4	2	1.11	0.438	1.27	0.5	1.905	0.75	2	1		4.5	20	0	0
20-Jul-81	4	11.9	39	2	4	2	0.64	0.25	1.111	0.438	1.746	0.688	1	1		4.75	24	6.35	2.5

TABLE 9--Database (Continued)

Date	Rod	Jt Spcng m	Jt Spcng (ft)	Sbstrt	Color	Brand	Center cm	Center (in.)	Bite cm	Bite (in.)	Width cm	Width (in.)	Orient.	Apprnc		Tacky	Hard	Pull cm	Pull (in.)
1	2	3	3	4	5	6	7	7	8	8	9	9	10	11	12	13	14	15	15
Polyurethanes (Continued)																			
20-Jul-81	4	11.9	39	2	4	2	0.64	0.25	1.111	0.438	1.746	0.688	1	1		4.75	24	6.35	2.5
31-Jul-81	4	6.1	20	2	3	2	0.32	0.125	0.476	0.188	3.175	1.25	1	3		5	22	12.7	5
31-Jul-81	4	6.1	20	2	3	2	0.64	0.25	0.635	0.25	1.905	0.75	2	1	4	5	31	6.35	2.5
06-Nov-81	4	7.62	25	2	3	2	0.48	0.19	1.27	0.5	1.905	0.75	2	3		4	23	10.2	4
06-Nov-81	4	7.62	25	2	3	2	0.24	0.09	1.27	0.5	1.905	0.75	1	5		4.5	34	7.62	3
01-Jan-82	2	6.71	22	2	3	22	0.64	0.25	1.27	0.5	2.223	0.875	3	5		5	55	2.54	1
01-Feb-82	3	0.61	2	2	3	2	1.27	0.5	1.27	0.5	0.635	0.25	1	5		4.5	25	0	0
01-Feb-82	4	1.22	4	2	3	2	0.95	0.375	1.27	0.5	1.27	0.5	3	1		3	28	2.54	1
01-Feb-82	4	1.22	4	2	3	2	1.59	0.625	1.588	0.625	1.905	0.75	2	5		4	12	5.08	2
02-Jul-82	4	7.32	24	2	5	2	0.64	0.25	1.27	0.5	0.794	0.313	4	5		4	30	5.08	2
02-Jul-82	4	7.32	24	2	5	2	0.64	0.25	0.794	0.313	0.794	0.313	1	5		4	31	3.81	1.5
15-Jul-82	3	2.13	7	2	3	2	0.08	0.031	0.079	0.031	1.905	0.75	2	1		4	18	8.89	3.5
15-Jul-82	3	1.22	4	4	3	2	0.08	0.031	0.079	0.031	1.588	0.625	2	6		1	0	0	0
23-Jul-82	2	6.1	20	3	6	2	0.48	0.188	0.953	0.375	1.905	0.75	4	5		4	41	5.08	2
23-Jul-82	2	6.1	20	3	6	2	0.48	0.188	1.032	0.406	1.588	0.625	3	1		4	41	0	0
23-Jul-82	3	6.1	20	2	3	8	0.64	0.25	0.794	0.313	2.223	0.875	1	5		3	25	7.62	3
30-Jul-82	4	7.62	25	1	4	2	0.32	0.125	1.032	0.406	2.54	1	2	1		3	21	3.81	1.5
30-Jul-82	4	7.62	25	1	4	2	0.56	0.219	1.111	0.438	1.429	0.563	3	1		2	22	7.62	3
08-Sep-82	4	3.66	12	1	3	2	0.79	0.313	0.953	0.375	0.794	0.313	5	6		1	11	10.2	4
29-Sep-82	4	7.32	24	2	3	2	0.64	0.25	0.953	0.375	1.588	0.625	1	1		4.5	22	7.62	3
29-Sep-82	4	7.32	24	2	3	2	0.64	0.25	0.635	0.25	0.953	0.375	4	1		4	21	7.62	3
29-Sep-82	4	2.44	8	2	3	15	0.16	0.063	0.318	0.125	0.318	0.125	5	5		5	55	6.35	2.5
29-Sep-82	3	6.1	20	2	5	2	0.87	0.344	0.873	0.344	0.794	0.313	4	1		4	15	7.62	3
06-Oct-82	4	6.71	22	2	3	2	0.64	0.25	0.953	0.375	2.54	1	2	1	3	5	20	2.54	1
06-Oct-82	4	6.71	22	2	3	2	0.64	0.25	0.476	0.188	1.27	0.5	1	5		5	20	7.62	3
11-Nov-82	5	13.7	45	1	5	2	0.64	0.25	0.635	0.25	1.588	0.625	2	1		4.5	17	6.35	2.5
30-Nov-82	4	2.44	8	4	2	2	0.64	0.25	0.635	0.25	0.635	0.25	3	3		4	25	3.18	1.25
30-Nov-82	4	2.44	8	4	2	2	0.64	0.25	0.635	0.25	0.635	0.25	1	1		4	29	3.18	1.25
30-Nov-82	4	2.44	8	4	2	2	0.95	0.375	1.27	0.5	1.27	0.5	2	5		4	25	1.27	0.5

TABLE 9--Database (Continued)

Date	Rod	Jt Spcng m	(ft)	Sbstrt	Color	Brand	Center cm	(in.)	Bite cm	(in.)	Width cm	(in.)	Orient.	Apprnc		Tacky	Hard	Pull cm	(in.)
1	2	3	3	4	5	6	7	7	8	8	9	9	10	11	12	13	14	15	15
Polyurethanes (Continued)																			
09-Dec-82	4	1.83	6	2	3	2	0.32	0.125	0.953	0.375	1.905	0.75	3	1		4.5	20	6.35	2.5
16-Dec-82	4	5.49	18	2	5	2	0.64	0.25	1.27	0.5	2.54	1	2	6		1	3	0	0
20-Dec-82	4	5.79	19	2	3	2	0.71	0.281	1.111	0.438	1.588	0.625	1	1		4	24	6.35	2.5
20-Dec-82	4	5.79	19	2	3	2	1.11	0.438	1.111	0.438	1.588	0.625	2	1		4	25	3.81	1.5
01-Jan-83	4	1.22	4	1	4	2	1.27	0.5	1.27	0.5	0.953	0.375	2	4		2	15	0	0
25-Jan-83	2	6.71	22	2	6	2	0.95	0.375	1.27	0.5	1.905	0.75	2	5		4	23	7.62	3
25-Jan-83	4	6.71	22	2	5	2	0.64	0.25	0.635	0.25	1.27	0.5	3	5		5	31	5.08	2
25-Jan-83	2	6.71	22	2	6	2	0.48	0.188	1.27	0.5	1.905	0.75	4	5		3	17	7.62	3
22-Apr-83	4	6.1	20	2	5	2	0.79	0.313	0.953	0.375	2.223	0.875	1	5		5	32	1.27	0.5
22-Apr-83	4	6.1	20	2	5	2	0.48	0.188	0.635	0.25	1.905	0.75	2	6		3	16	1.27	0.5
02-May-83	3	9.14	30	1	5	2	0.71	0.281	1.27	0.5	1.588	0.625	1	1		5	21	3.18	1.25
02-May-83	3	12.2	40	1	3	2	0.95	0.375	1.191	0.469	1.905	0.75	1	5		5	15	20.3	8
17-May-83	4	7.62	25	2	4	2	0.64	0.25	1.111	0.438	1.905	0.75	1	1		5	35	4.45	1.75
17-May-83	4	6.1	20	2	3	2	0.64	0.25	0.953	0.375	1.27	0.5	1	5		1.5	26	4.45	1.75
17-May-83	3	6.1	20	2	3	2	0.95	0.375	0.953	0.375	1.27	0.5	2	1	2	2.5	25	0	0
17-May-83	4	7.62	25	2	4	2	0.48	0.188	0.953	0.375	2.381	0.938	2	1		5	16	5.72	2.25
23-May-83	4	7.62	25	4	3	2	0.79	0.313	1.032	0.406	0.794	0.313	2	3		4	20	3.81	1.5
23-May-83	4	9.14	30	2	5	2	0.32	0.125	0.795	0.313	1.905	0.75	1	5		4	25	6.35	2.5
23-May-83	4	9.14	30	2	5	2	0.24	0.094	1.27	0.5	2.223	0.875	2	1		4.5	19	5.08	2
23-May-83	4	3.66	12	3	3	2	0.48	0.188	0.635	0.25	0.794	0.313	1	3		4	15	5.08	2
01-Jul-83	4	2.13	7	2	5	2	0.32	0.125	0.794	0.313	1.27	0.5	4	6		1	9	6.35	2.5
01-Jul-83	4	2.13	7	2	5	2	0.32	0.125	0.635	0.25	1.27	0.5	3	1		5	31	5.08	2
05-Sep-83	4	7.62	25	1	3	2	0.64	0.25	0.635	0.25	1.27	0.5	4	1		4	34	0	0
05-Sep-83	4	7.62	25	1	3	2	0.32	0.125	0.318	0.125	1.27	0.5	2	1		4	25	1.27	0.5
16-Sep-83	3	7.62	25	1	3	2	0.4	0.156	1.905	0.75	1.27	0.5	2	5		5	26	5.08	2
16-Sep-83	4	7.62	25	1	3	2	0.56	0.219	0.953	0.375	0.953	0.375	1	5		4	20	15.2	6
16-Sep-83	3	7.62	25	1	3	2	1.11	0.438	1.429	0.563	1.905	0.75	4	5		4	12	11.4	4.5
01-Oct-83	4	1.52	5	3	1	2	1.91	0.75	0.953	0.375	2.223	0.875	4	6		3	8	0	0
01-Oct-83	4	1.52	5	3	1	2	0.94	0.37	1.27	0.5	1.27	0.5	1	4		3	19	6.35	2.5

TABLE 9--Database (Continued)

Date	Rod	Jt Spcng m	(ft)	Sbstrt	Color	Brand	Center cm	(in.)	Bite cm	(in.)	Width cm	(in.)	Orient.	Apprnc		Tacky	Hard	Pull cm	(in.)
1	2	3	3	4	5	6	7	7	8	8	9	9	10	11	12	13	14	15	15
Polyurethanes (Continued)																			
01-Oct-83	4	1.52	5	3	1	2	0.16	0.063	0.476	0.188	0.953	0.375	2	6		3	15	3.81	1.5
24-Oct-83	4	6.1	20	2	3	2	0.32	0.125	0.635	0.25	1.27	0.5	2	1		4.5	25	1.27	0.5
24-Oct-83	4	6.1	20	2	3	2	0.32	0.125	0.476	0.188	0.635	0.25	1	5		5	45	8.89	3.5
01-Jan-84	4	0.91	3	1	2	2	0.64	0.25	0.635	0.250	1.27	0.500	2	6		1	0	0	0
01-Jan-84	3	2.44	8	2	3	22	0.95	0.375	1.27	0.5	0.635	0.25	4	1		5	40	4.45	1.75
01-Jan-84	4	5.49	18	2	3	22	0.95	0.375	1.27	0.5	1.905	0.75	5	4		3.5	18	10.2	4
01-Jan-84	4	0.46	1.5	1	2	2	0.79	0.313	0.953	0.375	1.27	0.500	1	5		4	24	3.81	1.5
01-Jan-84	4	5.49	18	2	3	22	0.64	0.25	1.27	0.5	1.588	0.625	4	1		4.5	25	3.81	1.5
01-Jan-84	4	5.49	18	2	3	22	0.95	0.375	1.27	0.5	1.905	0.75	4	5		2	13	17.8	7
01-Jan-84	4	0.76	2.5	1	2	2	0.32	0.125	0.635	0.250	1.27	0.500	1	1		3	26	6.35	2.5
09-Feb-84	4	3.66	12	2	5	2	0.56	0.219	0.953	0.375	0.635	0.25	2	1		5	25	6.35	2.5
09-Feb-84	4	3.66	12	2	5	2	0.95	0.375	1.588	0.625	2.223	0.875	1	5		5	32	3.81	1.5
29-Mar-84	4	6.1	20	2	4	2	0.48	0.188	0.794	0.313	1.588	0.625	2	1		4	26	4.45	1.75
29-Mar-84	4	6.1	20	2	4	2	0.79	0.313	0.953	0.375	0.953	0.375	1	1		4.5	22	5.72	2.25
14-Jul-84	3	5.49	18	2	3	22	1.59	0.625	1.588	0.625	2.223	0.875	2	6		2	5	5.08	2
14-Jul-84	4	1.22	4	1	3	22	0.64	0.25	0.953	0.375	1.27	0.5	2	6		3	19	5.08	2
15-Jul-84	4	3.05	10	1	4	2	0.48	0.188	0.476	0.188	1.27	0.5	4	6		1	5	0	0
15-Jul-84	4	3.05	10	1	4	2	1.59	0.625	1.588	0.625	1.905	0.75	4	5		5	50	2.54	1
15-Jul-84	4	4.57	15	2	5	2	0.32	0.125	0.635	0.25	0.794	0.313	1	5		5	32	5.08	2
15-Jul-84	4	4.57	15	2	5	2	0.48	0.188	1.27	0.5	0.794	0.313	2	1		5	32	5.08	2
22-Oct-84	4	7.62	25	2	5	2	0.64	0.25	0.794	0.313	1.27	0.5	1	5		5	25	5.08	2
22-Oct-84	4	7.62	25	2	5	2	0.64	0.25	0.953	0.375	1.588	0.625	2	1		5	22	3.81	1.5
01-Jan-85	3	2.44	8	2	3	22	1.27	0.5	0.635	0.25	0.965	0.38	2	5		5	22	0	0
01-Jan-85	3	1.83	6	2	3	22	0.64	0.25	0.483	0.19	1.27	0.5	2	1	2	1	12	0	0
01-Jan-85	4	7.62	25	2	6	22	0.41	0.16	1.118	0.44	1.118	0.44	1	1	2	4	25	0	0
01-Jan-85	4	6.1	20	2	6	22	0.32	0.125	1.113	0.438	2.54	1	2	6		1	9	0	0
01-Jan-85	2	6.1	20	2	6	22	1.11	0.438	1.27	0.5	1.905	0.75	1	6		1	18	7.62	3
01-Jan-85	3	2.44	8	1	3	22	0.64	0.25	0.965	0.38	0.787	0.31	2	4		5	60	0	0
01-Apr-85	4	7.32	24	1	5	2	0.48	0.188	0.476	0.188	1.27	0.5	1	5		4.5	23	7.62	3

TABLE 9--Database (Continued)

Date	Rod	Jt Spcng m	(ft)	Sbstrt	Color	Brand	Center cm	(in.)	Bite cm	(in.)	Width cm	(in.)	Orient.	Apprnc		Tacky	Hard	Pull cm	(in.)
1	2	3	3	4	5	6	7	7	8	8	9	9	10	11	12	13	14	15	15
Polyurethanes (Continued)																			
01-Apr-85	3	2.13	7	4	2	2	1.59	0.625	1.588	0.625	1.27	0.5	4	5		1	20	1.27	0.5
01-Apr-85	5	8.23	27	1	5	2	0.64	0.25	0.953	0.375	1.27	0.5	3	1		5	27	0	0
08-Apr-85	4	7.62	25	1	3	2	0.64	0.25	1.111	0.438	1.27	0.5	2	3		4.5	22	6.35	2.5
03-Jun-85	4	6.4	21	1	5	2	0.64	0.25	1.27	0.5	0.635	0.25	2	5		5	22	9.53	3.75
10-Jun-85	4	6.71	22	1	5	2	0.64	0.25	0.635	0.25	0.953	0.375	1	5		4.5	24	12.7	5
17-Jun-85	4	9.14	30	2	3	2	0.4	0.156	1.113	0.438	2.223	0.875	2	3		5	21	6.35	2.5
17-Jun-85	4	9.14	30	2	3	2	0.48	0.188	0.795	0.313	1.588	0.625	1	1		5	21	7.62	3
05-Aug-85	4	7.62	25	2	3	2	0.95	0.375	1.27	0.5	1.905	0.75	4	1		4	23	6.35	2.5
05-Aug-85	3	3.66	12	2	5	2	0.64	0.25	0.318	0.125	1.27	0.5	4	5		4	25	7.62	3
12-Aug-85	3	1.22	4	2	3	20	0.64	0.25	0.953	0.375	1.905	0.75	2	1	2	5	34	3.18	1.25
09-Sep-85	2	4.57	15	2	3	20	0.64	0.25	0.318	0.125	1.905	0.75	4	2		5	39	5.08	2
04-Nov-85	3	6.1	20	2	5	2	1.59	0.625	1.27	0.5	1.111	0.438	1	5		3	14	15.2	6
04-Nov-85	4	6.1	20	2	5	2	1.11	0.438	1.588	0.625	2.54	1	2	5		3	7	15.2	6
18-Nov-85	5	1.52	5	2	3	20	0.64	0.25	0.953	0.375	1.905	0.75	2	5		5	34	3.18	1.25
18-Nov-85	4	4.57	15	2	3	20	0.64	0.25	0.318	0.125	1.905	0.75	4	5		5	39	5.08	2
18-Nov-85	4	6.71	22	1	5	2	0.64	0.25	0.635	0.25	0.953	0.375	1	5		4.5	24	12.7	5
18-Nov-85	2	6.4	21	1	5	2	0.64	0.25	1.27	0.5	0.635	0.25	2	5		5	22	10.2	4
19-May-86	4	6.1	20	2	5	2	0.95	0.375	1.429	0.563	1.905	0.75	2	1		4.5	15	6.35	2.5
19-May-86	4	6.1	20	2	5	2	0.95	0.375	0.953	0.375	1.429	0.563	1	1		4.5	27	7.62	3
02-Jun-86	4	4.57	15	2	3	2	1.11	0.438	0.953	0.375	0.476	0.188	1	1		3.5	21	11.4	4.5
02-Jun-86	4	4.57	15	2	3	2	0.95	0.375	1.429	0.563	0.953	0.375	2	1		3.5	20	11.4	4.5
26-Sep-86	4	6.71	22	2	3	2	0.64	0.25	0.953	0.375	1.905	0.75	4	5		4.5	22	5.08	2
07-Oct-86	4	6.71	22	2	4	8	1.43	0.563	1.588	0.625	1.429	0.563	2	1		2	11	12.7	5
27-Oct-86	4	6.1	20	2	3	2	1.19	0.469	1.111	0.438	0.953	0.375	2	1		5	18	10.2	4
27-Oct-86	4	6.1	20	2	3	2	0.95	0.375	0.953	0.375	0.953	0.375	1	5		5	20	12.7	5
02-Mar-87	4	2.13	7	2	4	2	0.95	0.375	1.27	0.5	1.27	0.5	1	1		5	20	7.62	3
02-Mar-87	4	2.13	7	2	4	2	0.95	0.375	1.27	0.5	1.27	0.5	2	6		1	0	0	0
01-Jun-87	4	6.4	21	2	4	22	0.48	0.188	1.27	0.5	2.223	0.875	1	5		5	28	14	5.5
01-Jun-87	4	6.4	21	2	4	22	0.48	0.188	1.111	0.438	1.588	0.625	2	1		4	23	7.62	3

TABLE 9--Database (Continued)

Date	Rod	Jt Spcng m	(ft)	Sbstrt	Color	Brand	Center cm	(in.)	Bite cm	(in.)	Width cm	(in.)	Orient.	Apprnc		Tacky	Hard	Pull cm	(in.)
1	2	3	3	4	5	6	7	7	8	8	9	9	10	11	12	13	14	15	15
Polyurethanes (Continued)																			
31-Aug-87	5	7.62	25	2	3	8	0.79	0.313	0.794	0.313	1.905	0.75	3	2		5	35	10.2	4
31-Aug-87	4	5.49	18	2	5	8	0.95	0.373	0.953	0.375	1.588	0.625	1	5		5	21	6.35	2.5
02-Nov-87	4	6.1	20	3	4	2	0.64	0.25	1.27	0.5	2.54	1	4	5		5	26	0	0
02-Nov-87	4	6.1	20	3	4	2	0.64	0.25	1.27	0.5	4.445	1.75	4	1		5	19	0	0
02-Nov-87	4	6.1	20	3	4	2	0.79	0.313	1.588	0.625	2.54	1	4	1		5	34	0	0
08-Dec-87	4	6.71	22	2	3	8	0.79	0.313	1.27	0.5	2.223	0.875	1	5		4.5	19	22.9	9
08-Dec-87	4	6.71	22	2	3	8	0.48	0.188	1.27	0.5	1.905	0.75	2	5		4	15	17.8	7
03-May-88	4	6.1	20	2	3	8	0.64	0.25	1.746	0.688	2.064	0.813	1	5		3	12	17.8	7
03-May-88	4	6.1	20	2	3	8	0.32	0.125	1.111	0.438	1.905	0.75	2	5		3	19	8.89	3.5
08-Aug-88	4	7.62	25	2	3	8	0.95	0.373	0.953	0.375	1.588	0.625	1	5		5	21	6.35	2.5
08-Aug-88	4	7.62	25	2	3	8	0.95	0.375	1.113	0.438	1.27	0.5	3	5		4.5	20	6.35	2.5
12-Dec-88	4	6.1	20	2	3	8	0.79	0.313	1.27	0.5	1.27	0.5	2	5		3	22	5.08	2
12-Dec-88	4	6.1	20	2	6	8	0.48	0.188	1.27	0.5	2.223	0.875	1	5		3	18	8.26	3.25
12-Dec-88	4	6.1	20	2	3	8	0.79	0.313	1.27	0.5	1.588	0.625	1	5		4	23	5.72	2.25
12-Dec-88	4	6.1	20	2	6	8	0.48	0.188	0.953	0.375	1.27	0.5	2	3		3	17	6.35	2.5
01-Jan-89	2	2.44	8	2	5	22	0.64	0.25	0.635	0.25	1.27	0.5	2	6		2	8	0	0
01-Jan-89	2	2.44	8	2	5	22	0.79	0.31	1.27	0.5	1.905	0.75	1	6		1	5	0	0
01-Jan-89	3	2.44	8	2	3	22	1.27	0.5	1.27	0.5	1.27	0.5	1	2	1	5	55	0	0
01-Jan-89	4	6.1	20	2	3	22	0.48	0.188	0.953	0.375	3.175	1.25	1	2		3	15	0	0
08-May-89	4	1.83	6	4	2	7	0.79	0.313	0.794	0.313	0.635	0.25	4	5		4.5	17	4.45	1.75
08-May-89	4	1.83	6	1	5	7	0.64	0.25	0.794	0.313	1.588	0.625	1	5		5	25	5.08	2
01-Jul-89	2	6.71	22	2	3	14	0.64	0.25	1.27	0.5	3.81	1.5	5	4		5	32	6.35	2.5
01-Jul-89	2	6.71	22	2	3	14	0.32	0.125	0.953	0.375	3.175	1.25	4	2		5	52	6.35	2.5
04-Sep-89	3	9.14	30	2	3	8	0.79	0.313	0.953	0.375	1.905	0.75	1	2		5	26	1.27	0.5
04-Sep-89	2	9.14	30	1	3	9	0.64	0.25	0.953	0.375	1.27	0.5	4	5		4	18	0.64	0.25
04-Sep-89	3	9.14	30	2	3	8	1.27	0.5	0.953	0.375	1.905	0.75	4	2		5	31	0	0
04-Sep-89	2	9.14	30	1	3	9	0.48	0.188	0.953	0.375	1.429	0.563	4	5		4	20	2.54	1
18-Sep-89	1	3.05	10	2	6	8	1.11	0.438	1.111	0.438	1.111	0.438	2	3		4	33	2.54	1
18-Sep-89	4	4.27	14	2	1	9	0.64	0.25	0.953	0.375	1.111	0.438	1	5		5	29	7.62	3

TABLE 9--Database (Continued)

Date	Rod	Jt Spcng m	(ft)	Sbstrt	Color	Brand	Center cm	(in.)	Bite cm	(in.)	Width cm	(in.)	Orient.	Apprnc		Tacky	Hard	Pull cm	(in.)
1	2	3	3	4	5	6	7	7	8	8	9	9	10	11	12	13	14	15	15
Polyurethanes (Continued)																			
18-Sep-89	1	3.05	10	1	6	8	0.64	0.25	0.476	0.188	2.54	1	1	5		4.5	29	5.08	2
04-Dec-89	2	7.32	24	2	3	8	0.48	0.188	1.113	0.438	2.223	0.875	2	5		4	22	5.08	2
04-Dec-89	2	7.32	24	2	3	8	0.54	0.213	1.113	0.438	2.223	0.875	1	5		4	23	3.81	1.5
18-Dec-89	2	5.49	18	2	3	8	0.64	0.25	0.794	0.313	2.223	0.875	2	5		4.5	19	3.81	1.5
18-Dec-89	2	5.49	18	2	3	8	0.95	0.375	1.588	0.625	1.588	0.625	1	5		4.5	24	12.7	5
22-Jan-90	2	6.1	20	2	5	8	1.27	0.5	1.905	0.75	2.858	1.125	2	5		4	25	5.08	2
22-Jan-90	1	6.1	20	2	5	8	1.27	0.5	1.27	0.5	2.54	1	1	5		5	27	5.08	2
26-Feb-90	4	9.14	30	2	3	8	0.64	0.25	0.953	0.375	1.588	0.625	1	5		4.5	27	3.81	1.5
26-Feb-90	4	9.14	30	2	3	8	0.32	0.125	0.953	0.375	1.905	0.75	2	5		5	30	6.35	2.5
11-Feb-91	2	6.1	20	2	3	8	0.95	0.375	1.113	0.438	1.27	0.5	2	5		3	12	7.62	3
11-Feb-91	2	6.1	20	2	3	8	1.27	0.5	1.27	0.5	0.953	0.375	1	5		3	13	7.62	3
09-Oct-93	2	7.62	25	1	5	8	0.48	0.188	0.635	0.25	1.27	0.5	4	5		4	24	0.64	0.25
09-Oct-93	2	7.62	25	1	5	8	0.64	0.25	0.953	0.375	1.27	0.5	4	5		4	25	0.64	0.25

Michael A. Lacasse[1], Ralph M. Paroli[1]

EVALUATING THE DURABILITY OF SEALANTS

REFERENCE: Lacasse, M. A., and Paroli, R. M., **Evaluating the Durability of Sealants,"** Science and Technology of Building Seals, Sealants, Glazing and Waterproofing: Fourth Volume, ASTM STP 1243, David H. Nicastro, Ed., American Society for Testing and Materials, Philadelphia, 1995.

ABSTRACT:
Five silicone based sealants were subjected to heat aging, humidity and ultraviolet radiation over an eight thousand hour period. Changes in the tensile characteristics of the sealant products were determined and investigated on the molecular level using infrared and thermoanalytical techniques. Tensile testing revealed that specimens subjected to direct UV radiation were, in certain instances more susceptible to loss in elongation capacity than when shielded from radiation. Furthermore, the reduced elongation capacities were dependent on sealant type with the largest losses in the order of 53%. There were also noticeable changes in the modulus with decreases of 40-70% and for one product, an increase of 43% after 8000 hours of accelerated aging. It is surmised that modulus decreases are due to chain scission; however, it was not possible to strictly verify this from the analytical work carried out in this study. However, TG/DTA did confirm the susceptibility to loss of PDMS as well as other sealant components at elevated temperatures. PAS-FTIR was instrumental in detecting the presence of absorbed water as well as filler modifiers, such as calcium carbonate treated with stearic acid, which potentially play a role in the change in mechanical properties of aged silicone products.

KEY WORDS: sealant, sealant durability, sealant degradation, thermal analysis, durability, weathering, accelerated aging

1. Research Officers respectively, Materials Laboratory, Institute for Research in Construction, National Research Council Canada, Ottawa, ON, K1A 0R6.

INTRODUCTION

The degradation of sealants is due to a number of factors including exposure to elevated temperatures, moisture, ultraviolet (UV) radiation and cyclic movement. It is known that cyclic movement significantly affects the long-term performance of sealants [1,2]. However, the degree to which each of these factors contributes to the degradation process is less understood. To better understand the synergy between these degradation factors, materials are artificially aged in such a manner that these factors can be isolated and their impact assessed.

Weatherometer devices are often used to artificially age and accelerate the aging phenomena. Using these apparatuses, the intrinsic durability of sealants to a specific degradative action can be monitored in terms of changes in physical and chemical properties [3]. Changes in respect to physical phenomena may manifest themselves as, for example, variations in elongation capacity, strength, modulus, or similar mechanical characteristics.

Physical or chemical changes can also be investigated using analytical techniques such as thermal analysis (e.g., thermogravimetry (TG), differential thermal analysis (DTA), differential scanning calorimetry (DSC), simultaneous thermal analysis (STA), dynamic mechanical analysis (DMA) and thermomechanical analysis (TMA)) and spectroscopy (FTIR, NMR, MS, SEM, ESCA, etc.). These techniques have been successfully employed in numerous studies related to the degradation of polymeric based materials and can be used as a basis for investigating the synergistic effects of various combinations of aging factors.

Such types of analytical techniques are increasingly being used in the study of aging of sealants. Wolf [1] cites a study by Maurer and Brazier where thermomechanical analysis was used successfully to characterize the mechanical properties of sealant samples before and after aging. More recently Paroli et al. used thermoanalytical methods combined with spectroscopic analysis to investigate the aging process in roofing and sealant materials [4-8]. Analytical techniques are increasingly being relied upon to provide timely and cost effective results indicative of the extent of degradation in polymeric materials. Therefore, it is useful to review briefly some of the more salient aspects of the methods generally employed in this area.

Simultaneous Thermal Analysis

In STA, data from two thermoanalytical techniques such as thermogravimetry (TG) and differential thermal analysis (DTA), are acquired simultaneously (TG/DTA) by the same instrument on a single specimen. Since TG provides information on any reaction involving mass change as a function of temperature and/or time, information on the heat stability and composition of the sample can be obtained. Differential thermal analysis (DTA) provides information on the temperature difference between a reference material and a sample while they are subjected to a common temperature program.

Simultaneous TG/DTA technique is a very useful technique since physical and chemical changes of the sample are recorded simultaneously. In this respect it offers two particular advantages over the individual TG and DTA techniques: firstly, the problem of specimen homogeneity is eliminated since the two measurements are taken at the same time and temperature; secondly, it saves time since only one specimen has to be prepared. An additional benefit is that experimental errors are minimized since less sample handling takes place.

The STA technique is not always sufficient to characterize construction materials which have very complex compositions. In many cases, complimentary techniques (e.g., FTIR) are required to fully characterize the material.

Fourier Transform Photoacoustic Infrared (PAS-FTIR) Spectroscopy

Photoacoustic-FTIR spectroscopy has been used for a number of years in many different fields, including food, pharmaceutical, biomedical, coal, semi-conductors, environmental, and coatings. Although not as sensitive as some other infrared sampling techniques, PAS-FTIR has certain benefits [9,10]. Minimal sample preparation is a definite advantage of the PAS-FTIR technique. Photoacoustic spectroscopy renders possible the simultaneous study of sample and bulk properties. Moreover, the presence of up to 25 wt% carbon black, a highly absorbing filler used in many construction products, does not cause any significant interference. PAS-FTIR is particularly useful for measuring the spectra of materials, such as construction sealants, that suffer from a change in morphology due to weathering.

It is evident that a number of polymeric construction materials have very complex compositions. In the case of sealants, apart from the resin used as a modifier, plasticizers, stabilizers, fillers (carbon black and mineral) are also added. Weathered sealants cannot always be characterized by other commonly used IR techniques such as transmission or reflection IR, because they are black, gray or opaque, thick and, as previously mentioned, change morphology with weathering. PAS-FTIR technique is known to be capable of providing spectral information for this type of material with minimal sample preparation.

The object of this study was to compare the resistance of five silicone based sealants to accelerated aging in a xenon-arc weatherometer. Silicone sealants used in the construction industry offer particular advantages that include [11]: their one component nature, color stability, high temperature resistance (up to 200°C), great flexibility (down to about -60°C), UV and ozone resistance and good chemical resistance. These attributes are directly linked to their chemistry. These sealants contain polydimethylsiloxane rubber (PDMS) which has a very flexible polymeric backbone with high thermal stability [12]. This stability is due to the presence of very stable chemical bonds such as -Si-O-, -Si-C- and -Si-H. Furthermore, the polymer backbone is not affected by ultraviolet radiation, and therefore silicone sealants are less susceptible to this aging factor than other polymer based sealants. The flexibility of these polymer systems is brought about by a loosely crosslinked tri- or tetra-functional system that forms from a condensation curing reaction involving atmospheric moisture. There are a number of different moisture cured systems from which single component systems develop including: acid (acetoxy, octoate), neutral (bezamide, oxime, alkoxy) and alkaline (amine) cure systems [12]. The number of different types of moisture cured systems makes it possible to develop many types of products for specific applications.

Changes in mechanical properties (elasticity and strength) were evaluated using tensile tests, whereas change in chemical composition was monitored using PAS and STA. This work is part of a larger project which attempts to compare the weatherability of polyurethane, polysulfide and silicone sealants. The overall objective is to compare and correlate the various chemical and mechanical properties of these materials using natural and accelerated aging.

EXPERIMENTAL

Materials

Five commercially available silicone sealants were obtained from different manufacturers, a description of which is provided in Table I. Free film specimens were prepared by casting the sealant into brass forms that rested on a Teflon sheet, having inside dimensions of 102 mm x 152 mm x 3 mm (4" x 6" x 1/8"). The sealant was cured at standard conditions (23°C, 50% RH) for 6 hours after which the brass plate was removed; 10 days later, the free film was lifted from the Teflon sheet and then cut into three sections yielding films of 51 mm x 76 mm x 3 mm (2" x 3" x 1/8") which cured in the same standard conditions for another 20 days.

Table 1. Specimen Description

Silicone Sample	Color	Cure Mechanism	Movement* Capability	Hardness* (Shore A)
S1	Off-white	alkoxy	+100/-50	15
S2	White	alkoxy	±50	30
S3	Black	alkoxy	±50	n/a
S4	Gray	oxime	+100/-50	15
S5	White	methoxy	±50	22

* derived from manufacturer's literature

Procedure

Sample Weathering -- Two sections of the free films were placed back-to-back in a mounting frame, the films separated by an aluminum divider. This assembly was then placed into the xenon-arc weatherometer. One series of film was thus exposed to light and water spray, whereas the other was exposed to water only. The third section was kept in laboratory control conditions (23 °C, 50% RH) until testing. A set of samples was removed every 1000 hours and stored for further analysis. The total exposure was 8000 hours. The weathering conditions used for the samples were as follows:

Irradiance: 0.37 W m^{-2} nm
Black Panel Temperature: 63 °C
Relative Humidity Arc On: 50%
Light Cycle: 3.5 hours on; 0.5 hours off
Specimen Spray Cycle: 118 min. off; 20 min. on
Rack Spray Cycle: When arc off

Free Film Tensile Testing -- Dumbbell shaped sealant tensile test specimens (ASTM D412, die "C") were prepared from the exposed free-film samples in accordance with the prescribed test standard. Five specimens were cut from the water exposed films and five were cut from the water and light exposed films. All the samples were tested using a model 1122 Instron Universal testing machine equipped with an extensometer. A mean value for the various sealant properties was obtained with the aid of a data acquisition and analysis software package. The following settings were used:

Cross head Speed: 100 mm min^{-1}
Specimen Gauge Length: 10 mm
Extensometer Max. Displacement: 500 mm
Grip Spacing: 25 mm
Full Scale Load: 2 kg

Simultaneous Thermal Analysis -- TG/DTA studies of un-aged-weathered and weathered sealants were performed using a Seiko Simultaneous Thermal Analyzer (STA) model TG/DTA320. This module was interfaced to a Seiko SS5200H disk station. A specimen weighing between 7 and 8 mg was heated under a nitrogen atmosphere (100 mL/min.) from 40°C (no weight loss observed below 40°C) to 900°C at 10°C/min. All samples were run in duplicate.

Fourier Transform Infrared Photoacoustic Spectroscopy -- Fourier transform infrared spectra were measured on a Nicolet 170SX instrument equipped with an MTEC photoacoustic detector. For sampling, disks of 8 mm diameter were cut with a cork borer. In a few cases, where the samples were too thick, the thickness was reduced by slicing with a scalpel. Conditions used for recording the spectra were: resolution 8 cm^{-1}; mirror velocity setting 4 (0.1872 cm / s); number of scans 512 (21 min per spectrum); helium purge; Happ-Genzel apodization; standard phase correction. In addition to the weathered samples, two specimens of each material before exposure were analyzed and used as controls. In all cases, no significant difference was observed between the two specimens. All spectra were ratioed against a background recorded earlier the same day with carbon lampblack (Fisher Scientific, Cat. No. C-198) or with a carbon membrane standard reference. The spectra obtained were generally of good quality, but below 600 cm^{-1} only large features could be discerned because of the lower signal to noise ratios.

For most materials, spectra were recorded for unexposed samples and samples exposed for 2000, 4000 and 8000 hours.

RESULTS and DISCUSSION

Tensile Characteristics

The tensile characteristics of various silicone based sealants after weathering in an Atlas xenon-arc weatherometer are presented below in three parts. The first part offers a general review of the mechanical properties in tension of un-aged specimens, thus providing a means of categorizing the various types of sealants being evaluated. In an effort to appraise the effects of weathering, and hence provide a summary of results, un-aged specimens are compared to those aged for 8000 hours in a weatherometer. The second part results are reviewed on the basis of the phenomenological aspects related to changes in elongation and stress at rupture as a function of sealant type for products aged under conditions of direct UV radiation and elevated temperature. The third and final part provides an appreciation of the effect of the different aging regimes on the tensile properties of the sealant and highlights the relative importance of the various factors causing degradation in this study.

General Mechanical Characteristics -- The curves shown in Figures 1 to 7 depict the tensile characteristics of free film sealant specimens after 30 days curing at standard conditions. They are plotted in accordance with the theory of rubber elasticity in which the stress, σ, is related to the engineering strain by the following relation:

$$\sigma = G \bullet (\lambda - 1/\lambda^2)$$

where: G = shear modulus (MPa); λ = extension ratio = $1 + \varepsilon$; ε = engineering strain = $\Delta L / L_0$; and L_0 represents the initial specimen gauge length. Hence a plot of the stress as a function of the reduced strain function, λ^R, ($\lambda^R = \lambda - 1/\lambda^2$) provides a means of evaluating the shear modulus, G, which can readily be obtained from a portion of the straight line on the curve. These curves are representative of an averaged curve based on the testing of 4-5 specimens in tension and were developed by averaging values of stress and engineering strain at every point along the curve.

A review of the results of sealants tested prior to accelerated aging for the different products is provided in Figure 1. Essentially, these results confirm those values of Shore A hardness offered in the manufacturer's literature. Products S2 and S5 are characterized as high modulus sealants since both have Shore A values of 30 points (G_{S2} = 0.2 MPa; G_{S5} = 0.145 MPa), product S3 has an intermediate modulus (G_{S3} = 0.12 MPa), whereas products S1 and S4 are low modulus sealants having Shore A values of 15 points (G_{S1} = 0.04 MPa; G_{S4} = 0.025 MPa). No value was obtained for the Shore A hardness of product S3. As noted above, these values generally correspond reasonably well with those derived from the tensile tests.

As can be seen in Figure 2, variations in the shape of the stress-strain curves indicate an evident change in the tensile properties of most of the sealant products aged 8000 hours and exposed to UV radiation, elevated temperature and water spray. It is not evident from this plot, however, the extent to which changes occur in the initial stages of artificial aging nor those that occur over the course of the 8000 hour aging program.

Tensile Properties of Sealants as a Function of Aging Period -- Tables 2 and 3 provide values of the average stress and strain at rupture for the sealant products artificially aged for 4000 and 8000 hours in comparison to those not subjected to aging conditions. The representative stress-strain curves are given in Figures 3 to 7.

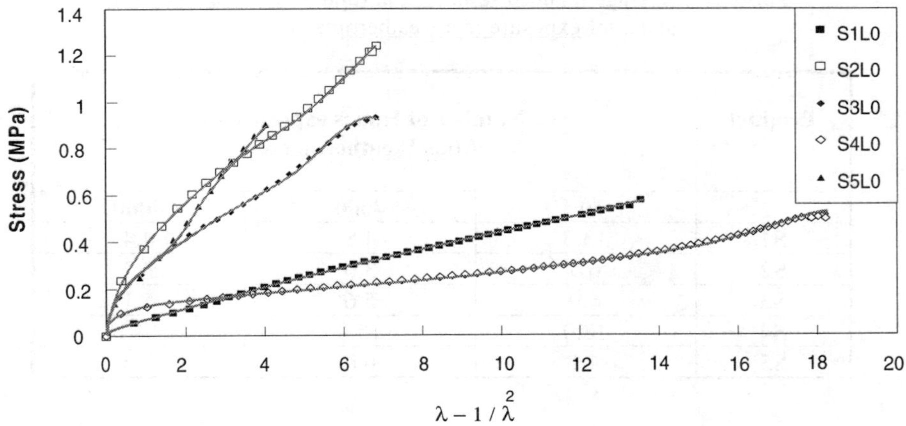

Figure 1. Stress as a function of λ - 1 / λ^2 for all materials in the unconditioned state (control samples).

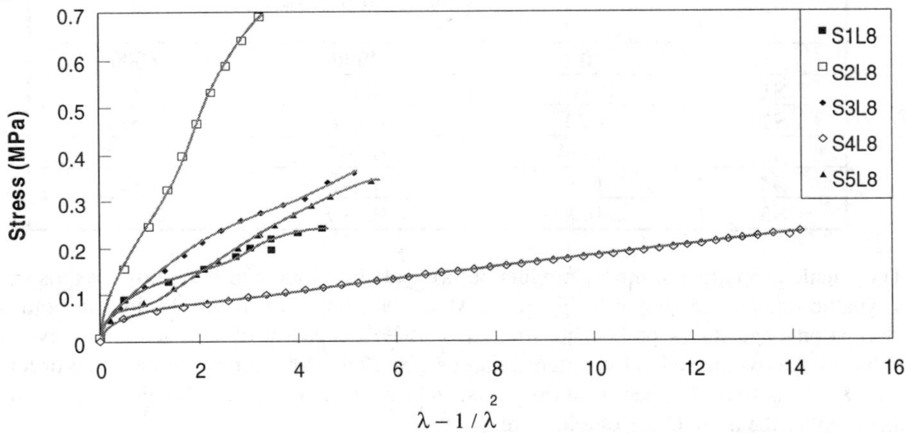

Figure 2. Stress as a function of λ - 1 / λ^2 for all materials conditioned for 8000 hours with UV light and water spray.

Table 2 -- Average reduced strain, λ^R, at rupture of sealants before and after exposure in a weatherometer

Product	Number of Hours exposure in Atlas Weatherometer		
	0	**4000**	**8000**
S1	13.5	4.8	4.4
S2	6.9	3.6	3.2
S3	6.9	5.6	5.1
S4	18.0	15.0	14.1
S5	4.1	6.0	5.5

Table 3 -- Average Stress (MPa) at Rupture, σ_r, of sealants before and after exposure in a weatherometer

Product	Number of Hours exposure in Atlas Weatherometer		
	0	**4000**	**8000**
S1	0.574	0.27	0.245
S2	1.24	0.678	0.678
S3	0.924	0.387	0.352
S4	0.5	0.248	0.227
S5	0.9	0.410	0.335

The tensile properties of the S1 product before and after 4000 and 8000 hours exposure in a weatherometer are shown in Figure 3. After 4000 hours exposure the loss in both the elongation capacity (- 65%) and strength (- 53%) capacity of this sealant is evident, although the overall reduction in tensile properties after 8000 hours is in the same order of magnitude as that obtained at 4000 hours, indicative of the minor changes in properties afforded by the extra hours of exposure.

Figure 4 shows the stress-strain characteristics of the S2 product, where it is observed that there is also a reduction in tensile properties after 4000 hours ($\Delta\lambda^R$ = - 47%; $\Delta\sigma_r$ = - 45%) and 8000 hours ($\Delta\lambda^R$ = - 53%; $\Delta\sigma_r$ = - 45%). The change in properties after 4000 hours is not as significant, suggesting that most of the changes occur before the 4000 hour aging period.

These results are similar to those obtained in the case of the S3 product, as shown in Figure 5. Loss in elongation capacity after 4000 and 8000 hours of aging are essentially the same (i.e.$\Delta\lambda^R$ @ 4000 hrs = - 18%; @ 8000 hrs = - 25%) as is the loss in strength for the same aging conditions (i.e. $\Delta\sigma_r$ = - 58% & - 62% respectively). The

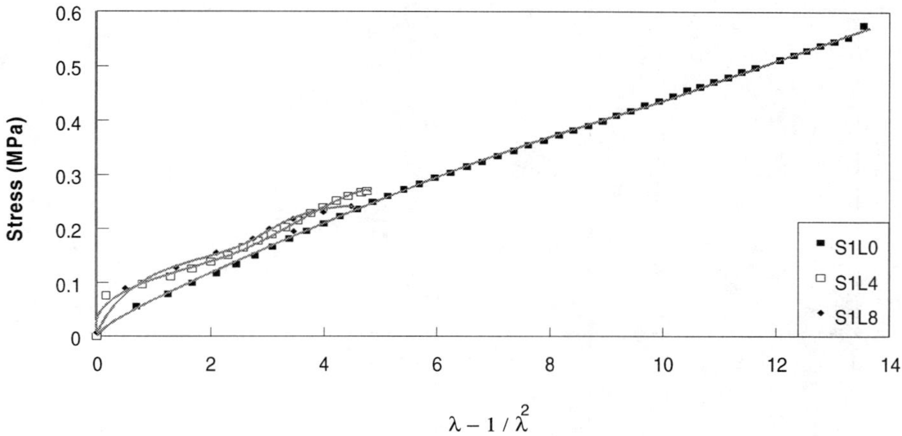

Figure 3. Stress as a function of $\lambda - 1/\lambda^2$ for S1 conditioned in UV light and water spray for 0 ,4000, and 8000 hours.

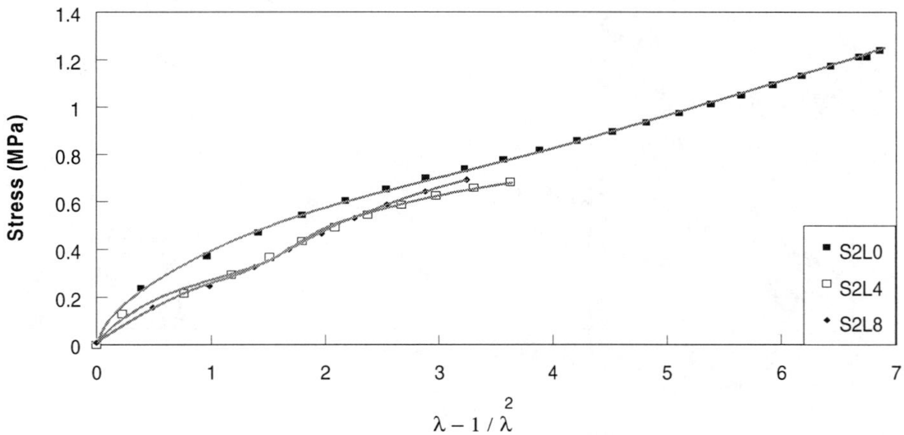

Figure 4. Stress as a function of $\lambda - 1/\lambda^2$ for S2 conditioned in UV light and water spray for 0 ,4000, and 8000 hours.

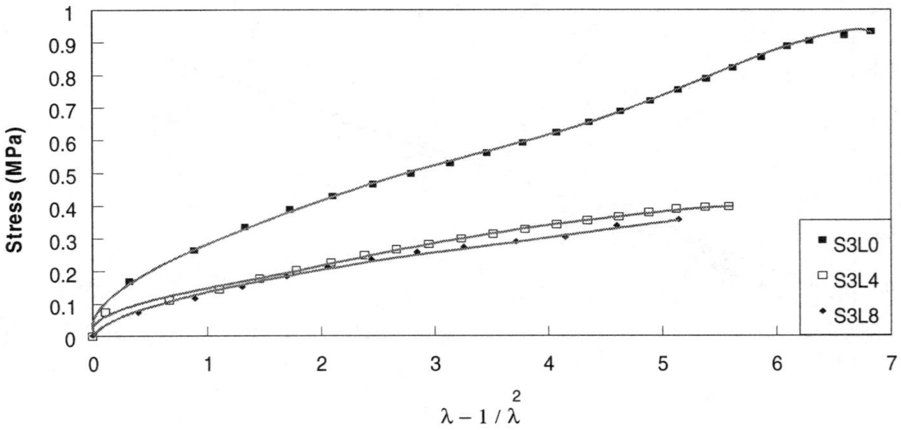

Figure 5. Stress as a function of $\lambda - 1 / \lambda^2$ for S3 conditioned in UV light and water spray for 0 ,4000, and 8000 hours.

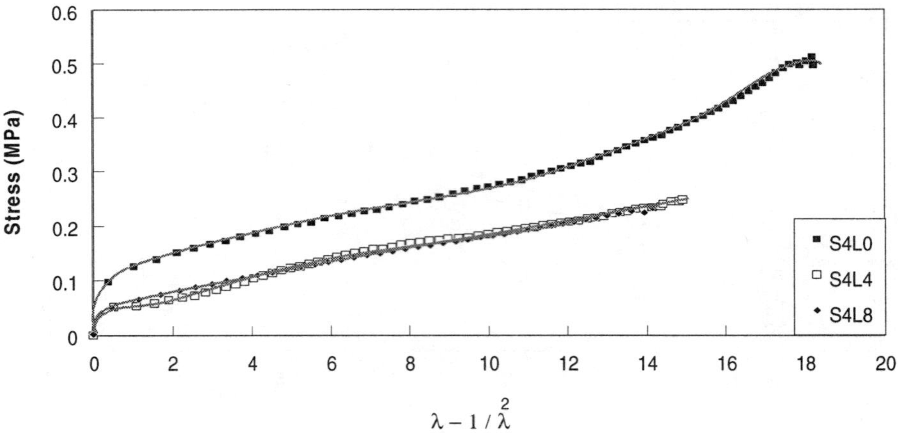

Figure 6. Stress as a function of $\lambda - 1 / \lambda^2$ for S4 conditioned in UV light and water spray for 0 ,4000, and 8000 hours.

more notable change is that of the tensile shear modulus given by the slope of the stress-strain curve along the linear portion of the curve. The loss in modulus may be due to chain scission. Loss in elongation capacity is most probably a result of the increased incidence of micro defects at the surface and edges of the sealant specimens, which tend to increase the probability of premature failure of the specimen.

The product S4 has the greatest elongation capacity (Figure 6) before being subjected to accelerated aging and it retains a significant capability to stretch even after prolonged exposure to UV radiation (i.e., $\Delta\lambda^R$ = -16% & 22% @ 4000 & 8000 hours exposure respectively.). More significant changes occur with respect to the rupture stress (i.e. $\Delta\sigma_r$ = - 50% & - 55% @ 4000 & 8000 hours exposure respectively). However, the apparent change in tensile modulus is slight, suggesting a product which is particularly resistant to accelerated aging under those conditions imposed in this study.

The most significant property change for product S5 is that of the tensile modulus. As seen in Figure 7, not only does accelerated aging produce a reduction in values of modulus after both 4000 and 8000 hours of exposure, but also there is an increase in the elongation capacity of the product ($\Delta\lambda^R$ = + 49% & + 35% @ 4000 & 8000 hours exposure respectively). The reduced modulus and increased elongation capacity necessarily causes a decrease in the value of the stress at rupture ($\Delta\sigma_r$ = - 54% & - 63% @ 4000 & 8000 hours exposure respectively). The response to aging of the S5 product is similar to that observed for the S3 product, both of which are structural glazing sealants. Evidently, for these types of sealants, the loss in modulus is not as important to the performance of the product as is an increased incidence of failure manifesting itself as a reduced elongation capacity.

Figure 8 shows the change in shear modulus of sealant products as a function of exposure time over the entire aging program. No change in modulus is evident for product S1; however, product S2 tends to produce a net increase in modulus (ΔG = + 43%) indicating a possible crosslinking action brought about by the aging process. A more significant loss in modulus is seen for product S3 (ΔG = - 58%) and a similar trend is observed for product S4 (ΔG = - 40%). Product S5 has the larger loss in modulus (ΔG = -70%) of all the products evaluated.

Another aspect that can be observed from these results is that changes occur early on in the aging program, with most products providing evident changes in modulus after 2000 hours exposure and all products after 3000 hours. This suggests that extended periods of exposure up to 8000 hours duration need not necessarily be required to achieve an indication of a products' resistance to accelerated aging. On the other hand, long-term trends are difficult to predict without extensive and thorough testing programs.

Comparison of Aging Methods -- In order to ascertain the degree to which mechanical properties are affected by the different aging programs, a comparison is provided between each method using the change in elongation capacity as the point of comparison. Values for the different products subjected to aging conditions within the weatherometer that include exposure to UV radiation and those that do not are provided in Figures 9 and 10. Changes in elongation capacity as a function of the control value for products exposed and not exposed to UV radiation are given in Table 4. These values represent the ratio between results of samples not exposed to those exposed to direct UV radiation.

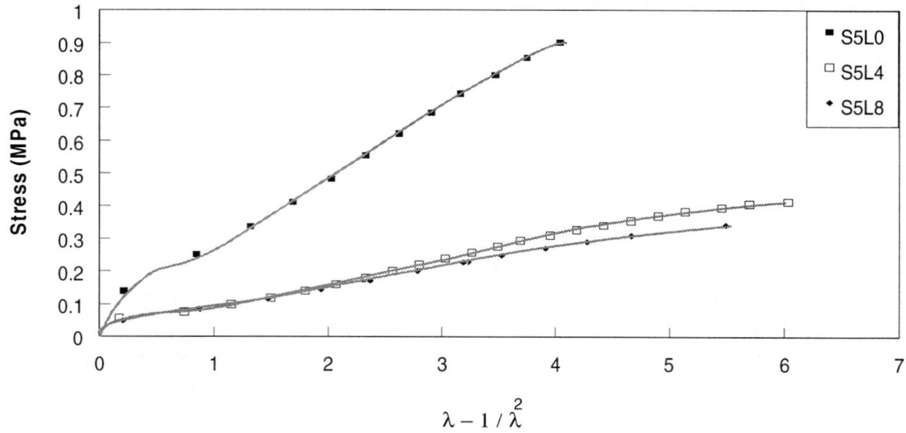

Figure 7. Stress as a function of $\lambda - 1 / \lambda^2$ for S5 conditioned with UV light and water spray for 0 ,4000, and 8000 hours.

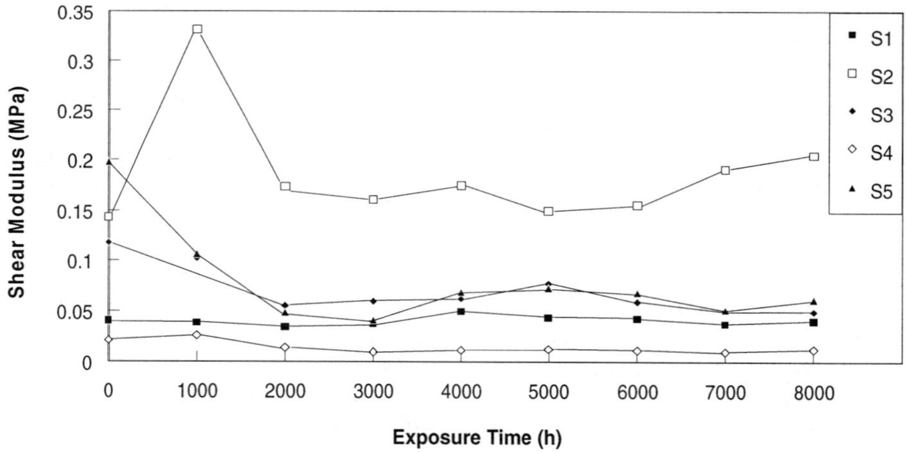

Figure 8. Shear Modulus as a function of conditioning time when exposed to UV light and water spray for each of the sealants.

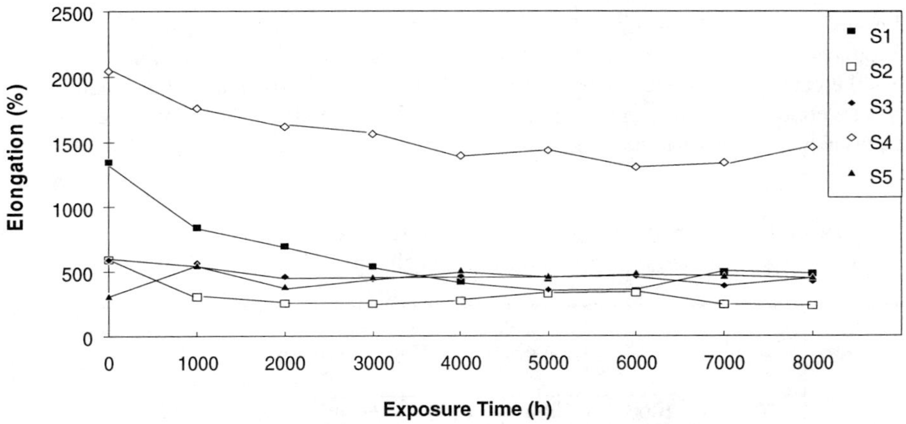

Figure 9. Elongation at break as a function of conditioning time when exposed to UV light and water spray for each of the sealants.

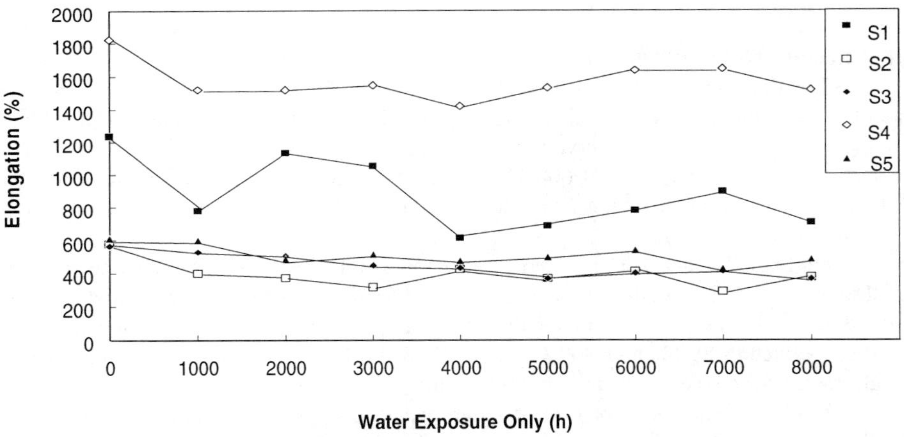

Figure 10. Elongation at break as a function of conditioning time when exposed to water spray only for each of the sealants.

It is apparent from an examination of the values in Table 4 that exposure to UV radiation reduces, in all cases, the elongation capacity of the sealants. This is more pronounced for products S1 and S2 which, on average, provide approximately 1.5 to 2 times the capacity to stretch when not exposed to direct UV radiation than under direct exposure conditions. The remaining products have less important reductions in elongation capacity when comparing either exposure regime.

Table 4. -- Ratio of Tensile Elongation at Rupture of Non-Exposed to that of exposed Sealant Specimens to Direct UV Radiation

Product	Number of Hours exposure							
	1000	**2000**	**3000**	**4000**	**5000**	**6000**	**7000**	**8000**
S1	1.07	1.75	2.02	1.58	1.96	2.33	2.32	2.04
S2	1.41	1.42	1.26	1.66	1.15	1.39	1.22	1.76
S3	1.09	1.22	1.04	0.99	1.1	1.04	1.30	0.86
S4	0.94	0.99	1.08	1.04	1.12	1.25	1.18	1.08
S5	1.15	1.23	1.12	1.00	1.14	1.19	0.93	1.1

Simultaneous Thermal Analysis

STA data provided simultaneous DTA and TG/DTG results of the silicone sealants. A typical STA plot (TG, DTG and DTA) is shown in Figure. 11a. Two endothermic events can be observed in the DTA curve shown in this figure. The first (between approximately 300 and 600°C) is related to the decomposition of the polymer while the second event (between 700 and 800°C) is most likely due to the decomposition of the filler (e.g., $CaCO_3$).

The TG/DTG analysis revealed two major weight losses for the silicone sealants (see Figure 11b). The first weight loss occurred in the 200-685°C region and is most likely due to the decomposition of the siloxane polymer. The second weight loss, observed between 685°C and 800°C, arose from the breakdown of calcium carbonate. Weight losses for the sealants at 0 and 6000 hours are summarized in Tables 5.

The S1 sealant does not undergo significant changes with weathering (see Table 5). The samples start to lose weight at ~300 °C and the total weight loss is 67-69% for both the unexposed and UV/water exposed samples. This variation may be attributed to inhomogeneity in the sealant rather than to weathering.

The S2 sealant series has a total weight loss of 64-65 %. The S2 series started to lose weight at a lower temperature than the S1 samples (~200°C). Hence, it is thermally less stable. The TG and the DTG curves of the S2 samples displayed similar trends to those observed for S1 sealants. However, the peak temperature of the two weight losses for the samples exposed to light and/or water were lower than that of the control. Therefore, the weathering appears to affect thermal stability.

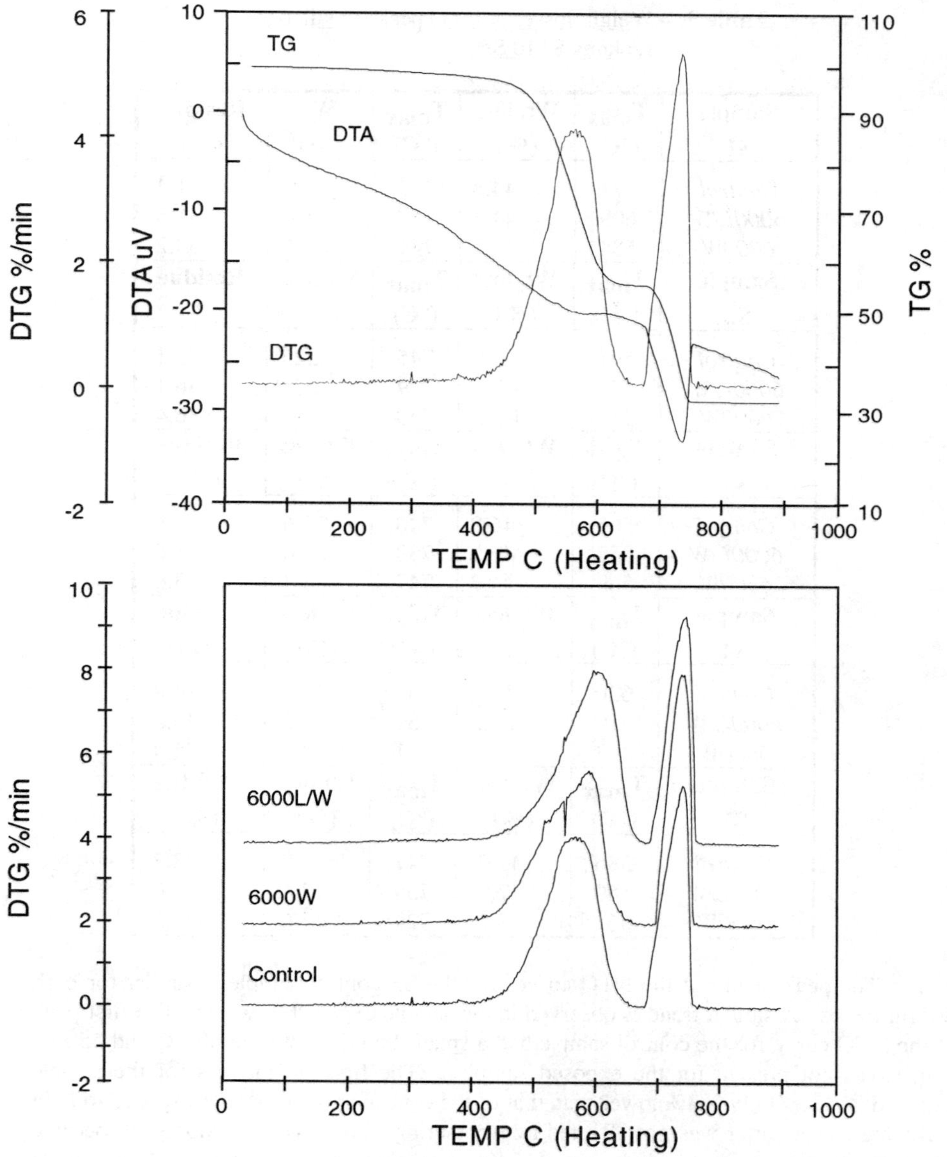

Figure 11. a) Typical Simultaneous Thermal Analysis (STA) plot showing TG, DTG and DTA as a function of temperature. b) DTG curve for silicone sealant specimens subjected to 6000 hours of accelerated aging: 6000 L/W represents plot of specimens exposed to direct UV radiation and water spray; 6000W specimens shielded from direct radiation; control is the un-aged specimen.

Table 5. -- Weight losses at DTG peak for silicone sealants S1 to S5

Sample S1	T_{max} (°C)	Wt. loss (%)	T_{max} (°C)	Wt (%)	Residue (%)
Control	571	44.8	743	23.8	31.4
6000L/W	609	44.3	749	24.2	31.5
6000W	584	42.5	741	24.4	33.2
Sample S2	**T_{max} (°C)**	**Wt. loss (%)**	**T_{max} (°C)**	**Wt loss (%)**	**Residue (%)**
Control	591	42.0	745	22.9	35.1
6000L/W	554	40.9	739	23.0	36.1
6000W	573	40.7	740	22.9	36.4
Sample S3	**T_{max} (°C)**	**Wt. loss (%)**	**T_{max} (°C)**	**Wt loss (%)**	**Residue (%)**
Control	565	46.9	740	21.6	31.6
6000L/W	555	46.9	738	21.6	31.6
6000W	569	44.3	747	22.7	33.0
Sample S4	**T_{max} (°C)**	**Wt. loss (%)**	**T_{max} (°C)**	**Wt loss (%)**	**Residue (%)**
Control	620	56.1	750	19.1	24.8
6000L/W	619	55.5	753	19.3	25.3
6000W	593	54.8	741	19.3	25.9
Sample S5	**T_{max} (°C)**	**Wt. loss (%)**	**T_{max} (°C)**	**Wt loss (%)**	**Residue (%)**
Control	586	47.0	747	17.9	33.7
6000L/W	560	47.9	739	17.4	35.0
6000W	561	48.2	729	17.6	35.3

The peak height of the DTG curves for the S3 control sample is similar for both weight losses. A similar trend is observed in the sample exposed to water. The first peak of the DTG curve for the control sample had a small shoulder between 400°C and 550°C, which was not present for the exposed samples. The first weight loss for the sample exposed to water only is 3% lower than that of the control and the sample exposed to light and water. The latter was not affected by weathering; however, the same peak became sharper and an increase in peak height of ~ 40% was observed. Although only the sample exposed to water shows a 3% lower weight loss, the shoulder in the control sample was not observed in either of the exposed samples. This may be due to the loss of a component in the sample during exposure to water. The sharpness of the first derivative weight loss of the DTG curve for the sample exposed to light, as well as its increase in peak height, shows that this sample is thermally less stable.

The DTG curve for the S4 control silicone sample revealed that the sample lost weight at a similar temperature (~ 300°C) as the S1 and S3 series. The S4 series showed the largest total weight loss (75%) of all series studied, and the control and exposed samples showed similar weight losses. A comparison of the weight losses for this sample with those for the previous samples reveals that S4 loses more weight in the first region and less in the second. The peaks of the DTG curves for the control sample in both weight loss regions have similar peak heights. UV radiation and/or water exposure seem to have an opposite effect on the shape of the peaks of the DTG curves for this series. Both exposed samples showed two unresolved peaks for the first weight loss, but this peak becomes sharper for the sample exposed to water. In addition, the peak height of the second peak decreased slightly for the sample exposed to water, whereas it became sharper for the sample exposed to light. Although the DTG curves for the control and exposed samples have different shapes, the series does not seem to be significantly affected by the two types of weathering conditions used. However, it is more affected than S1, S2, and S3.

The TG/DTG curves for control and exposed S5 silicone series are summarized in Table 5. A shoulder between 250°C and 500°C was observed for the control sample. The first DTG peak for the control sample was higher than the second one. Samples exposed to UV-radiation and/or water did not show the shoulder on the first peak and the two weight losses occurred at temperatures slightly higher than that of the control sample. They also showed a slight decrease in the peak height of the first peak and with a proportional increase on the peak height of the second peak.

FTIR Analysis

Generally, there was very little change in the FTIR spectra for the five different sealants after weathering. Representative FTIR spectra of a silicone sealant (before and after weathering) is shown in Figure 12. All of the sealant spectra contained the general peaks representative of the siloxane polymer. For example, the C-H stretching peaks at 2966 and 2904 cm^{-1}, the broad Si-O-Si band at 1200-1000 cm^{-1}, the Si-CH_3 peaks at 1267 and 822 cm^{-1}, and the peak at 700-660 cm^{-1} [8]. Furthermore, a number of peaks typical of calcite (a type of calcium carbonate) were observed: a broad band from 1600 to 1300 cm^{-1}, sharper peaks at 879 and 712 cm^{-1}, and weak peaks at 2513 and 1796 cm^{-1}.

The spectrum of the unexposed S1 specimen resembled that shown in Figure 12. The only significant difference after weathering was the appearance of a broad OH band near 3400 cm^{-1}, which could be due to absorbed water. The spectrum of the control S2 sample shows the same peaks observed in S1, corresponding to calcium carbonate and polydimethylsiloxane. The shape of the carbonate band at 1600-1300 cm^{-1} was slightly different, which may indicate the use of a different type of precipitated form of calcium carbonate. As for S1, weathering gave rise to the appearance of a broad OH band near 3400 cm^{-1}. In addition, the sample exposed to water for 2000 hours displayed peaks at 2919, 2850, and 1579 cm^{-1}. The first two peaks are typical of long hydrocarbon -$(CH_2)_n$- chains, and the third could be due to carboxylate ions -COO-. Peaks at these positions are characteristic of calcium salts of fatty acids such as calcium stearate. This was not due to contamination. Calcium carbonate is available in a form treated with stearic acid. It is

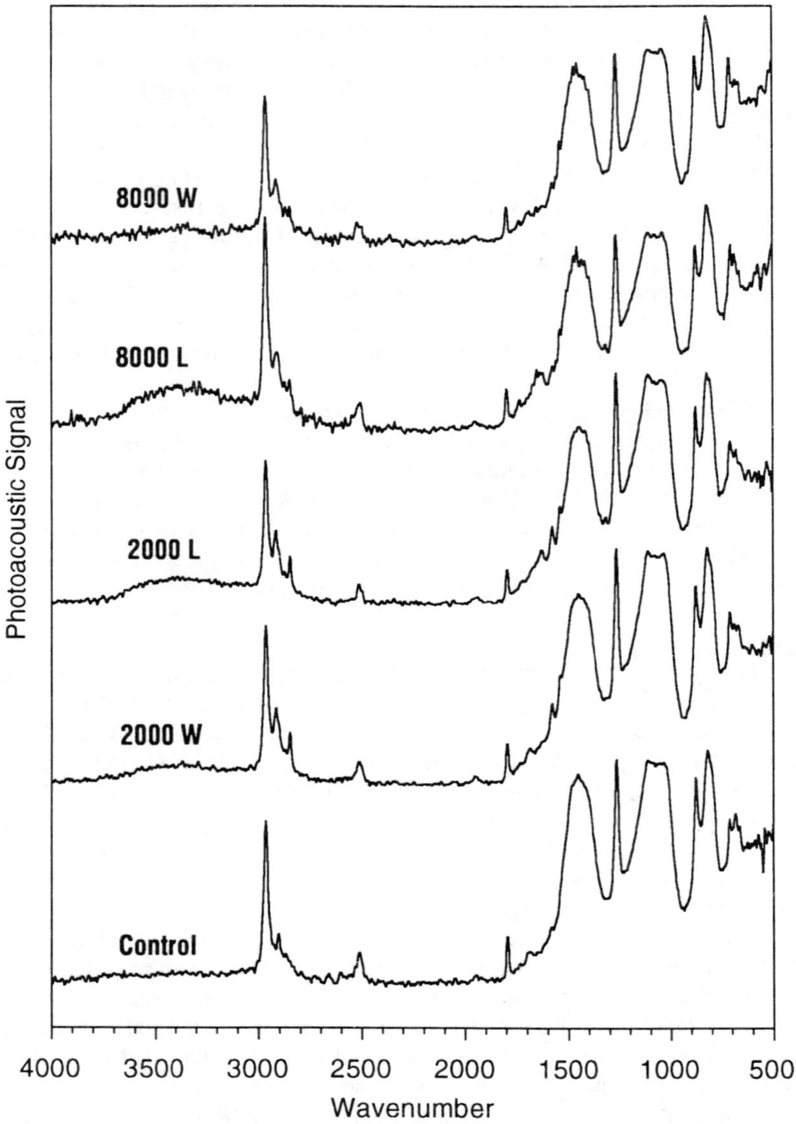

Figure 12. Representative FTIR spectra of a silicone sealant before (control spectra) and after accelerated aging.

possible that this type of calcium carbonate was used and that the weathering resulted in surface enrichment in stearate.

Unlike the previous two materials, the S3 sealant was black while the S4 sealant was gray. The S3 sample was loaded with carbon black, and this probably accounts for the higher baseline observed in the spectra (about 5% as compared to less than 1% in the previous cases). Otherwise, the spectra show the same peaks as S1, arising from calcium carbonate and polydimethylsiloxane. Again, exposure to water or UV radiation/water spray for 2000, 4000 and 8000 hours has little effect on the spectrum. The increase in OH absorption on exposure is less pronounced than seen in S1 and S2. Lastly, as in the S2 case, the calcium stearate peaks were observed.

The spectra for the S5 sealant also consisted of calcium carbonate and polydimethylsiloxane peaks. However, a weak shoulder near 1690 cm^{-1} was also observed. It would also appear that another pigment such as titanium dioxide was used, since higher absorption in the 700-500 cm^{-1} range was noted. The shape of the carbonate band is similar to that seen in S2. The samples exposed to water and UV radiation/water spray showed an increase in OH absorption near 3400 cm^{-1}. It is interesting to note that the exposed S5 samples also contained peaks characteristic of calcium stearate at 2918, 2850, 1578, 1540 cm^{-1}.

SUMMARY

Based on the tensile testing of these sealant products it is apparent that they are capable of sustaining significant extension to failure. The inherent flexibility of these materials is retained even after being subjected to extensive accelerated aging. However, the tensile characteristics after aging do vary according to the sealant type, with product S4 showing the least change in properties after 8000 hours of aging. The elongation capacity of products S1 and S2 is particularly affected by accelerated aging, although these products still retain a considerable capacity to stretch after prolonged weathering. The product S2 also shows an increase in modulus after aging, indicative of a possible crosslinking effect. Loss in modulus is characteristic of products S3 and S5, most probably due to chain scission. A comparison of different aging methods reveals that exposure to UV radiation in conditions imposed in this study reduces, in all cases, the elongation capacity of the sealants.

Thermoanalytical studies undertaken in this investigation indicate that the main weight losses observed by TG/DTA are most likely due to the siloxane polymer and calcium carbonate. These weight losses are characteristic of chain scission which manifests itself as a loss in modulus and a corresponding reduction in elongation capacity. Additional work using a hyphenated technique such as TG/DTA-FTIR or quenching the sample at various temperatures would facilitate the identification of the various degradation components.

Spectroscopic analysis revealed that the five silicone-based materials evaluated all have a similar composition. The main components are a polydimethylsiloxane polymer and calcium carbonate; the S5 product may also contain some titanium dioxide. Only minor changes were observed using FTIR-PAS spectroscopy of the five materials exposed to 8000 hours of aging. In some cases there is an increase in hydroxyl absorption (possibly signifying an increase in water absorption). Furthermore, the weathering of specimens S2 and S5 resulted in surface enrichment in calcium stearate.

REFERENCES

1. Wolf, A., "Studies of the Aging Behavior of Gun-grade Building Joint Sealants - State-of-the-Art", Polymer Degradation and Stability, Vol. 23, 1989, pp. 35-163.

2. Lacasse, M.A., "Advances in Test Methods to Assess the Long-Term Performance of Sealants", Science and Technology of Building Seals, Sealants, Glazing and Waterproofing: 3rd Volume, ASTM STP 1254, James C. Myers, Ed., American Society for Testing and Materials, Philadelphia, (1994), pp. 5-20.

3. Wypych, J., Weathering Handbook, Chemtec Publishing, Toronto, 1990.

4. Paroli, R.M., Dutt, O., Delgado, A.H., Mech, M.N., Thermochimica Acta, Vol. 182, No. 191, 1992, pp. 303-317.

5. Paroli, R.M., Dutt, O., Delgado, A.H., Stenman, H.K., Journal of Materials in Civil Engineering, Vol. 5 No. 1, 1993.

6. Paroli, R.M., Delgado, A.H., "Applications of TG-FTIR in the Characterization of Weathered Sealants" American Chemical Society, Division of Polymeric Materials:Science and Engineering, Vol. 68, 1993, pp. 139-140.

7. Paroli, R.M., Delgado, A.H., "Applications of Thermogravimetry in the Characterization of Silicone Sealants" American Chemical Society, Division of Polymeric Materials:Science and Engineering, Vol. 68, 1993, pp. 334-335.

8. Paroli, R.M., Delgado, A.H., Cole K., "Applications of thermogravimetry and PAS-FTIR in the Characterization of Silicone Sealants", Canadian Journal of Applied Spectroscopy, Vol. 39, No. 1, 1993, pp. 7-14.

9. Vidrine, D.W., "Photoacoustic Fourier Transform Infrared Spectroscopy of Solids and Liquids", Chapter 4 (pages 125-148) in: Fourier Transform Infrared Spectroscopy. Vol. 3, J.R. Ferraro and L.J. Basile (Eds.), Academic Press, 1982.

10. Graham, J.A., Grim III, W.M. and Fately, W.G., "Fourier Transform Infrared Photoacoustic Spectroscopy of Condensed-Phase Samples", Chapter 9 (pages 345-392) in : Fourier Transform Infrared Spectroscopy, Vol. 4, J.R. Ferraro and L.J. Basile (Eds.), Academic Press, 1985.

11. Panek, J.R., Cook, J.P., Construction Sealants & Adhesives, 3rd edition, John Wiley & Sons, New York, N.Y., 1991, pp. 118-128.

12. Wolf, A., "Material Properties of Construction Sealants", Kautschuk & Gummi Kunststoffe, Vol. 41, No. 2, 1988, pp. 173-178.

Michael A. Lacasse[1], Jennifer E. Bryce[1], James C. Margeson[1]

EVALUATION OF CYCLIC FATIGUE AS A MEANS OF ASSESSING THE PERFORMANCE OF CONSTRUCTION JOINT SEALANTS: SILICONE SEALANTS

REFERENCE: Lacasse, M. A., Bryce, J. E., and Margeson, J. C., **"Evaluation of Cyclic Fatigue as a Means of Assessing the Performance of Construction Joint Sealants: Silicone Sealants,"** Science and Technology of Building Seals, Sealants, Glazing, and Waterproofing: Fourth Volume, ASTM STP 1243, David H. Nicastro, Ed., American Society for Testing and Materials, Philadelphia, 1995.

ABSTRACT: Five types of silicone based high performance construction sealant were subjected to cyclical movements using manually adjusted vices until rupture of the specimens occurred. Failure of these sealant specimens during cyclic fatigue testing allowed comparison with previously obtained results for silicone sealants. Sealant fatigue behavior was compared with the fatigue behavior of elastomers and rubbers in order to gain a more quantitative understanding of sealants subjected to fatigue testing. These results were used to give an assessment of sealant performance in relation to their dynamic fatigue life based upon sealant resistance to cyclic deformation.

KEYWORDS: sealants, silicone, polyurethane, polysulphide, cyclic fatigue testing, sealant fatigue behavior, sealant service-life performance,

1. Research Officer, laboratory assistant, and technical officer, respectively, National Research Council Canada, Institute for Research in Construction, Materials Laboratory, Ottawa, Ontario, Canada, K1A 0R6;

INTRODUCTION

Sealants are elastomeric based compounds characterized not only by their rubber-like flexibility, but also by their ability to effectively seal construction joints against the elements, preventing the passage of air, water, and debris. This inherent flexibility enables them to respond to and withstand repeated cycles of extension and compression as imposed on them by adjacent building elements. The cyclical joint movements are in response to both daily and seasonal temperature fluctuations that cause the expansion and contraction of building elements adjacent to the joint thus closing the joint as temperature rises and opening it as temperature falls. Hence the joint sealant undergoes daily fatigue cycle superimposed on a yearly cycle of reversible joint movement. Consequently, the dynamic fatigue life of sealant in a joint is an important material characteristic that must be taken into consideration when selecting a sealant since the improper choice of product may result in premature repair or replacement of the sealed joint. This is often a very costly task.

Previous investigations by Karpati [1,2] have shown that cycling is the most important service factor influencing sealant performance. Much of this work involved cyclic fatigue tests of various sealants both indoors and out. Correlation between outdoor performance and laboratory fatigue testing of sealant specimens has been demonstrated in Karpati's earlier work on a two-part polysulfide sealant [3]. In this investigation, fatigue testing in manually-adjusted vises in the laboratory allowed actual time to failure to be accelerated, but the number of cycles to each mode of failure was consistent with those of sealant specimens exposed and cycled on a strain-cycling exposure rack [4], an apparatus simulating building joint movement.

The method devised by Karpati for fatigue testing in the laboratory has also been applied to both polyurethane and polysulfide sealants [5,6] at different rates of cyclic deformation, thus suggesting that it may be widely applicable to other types of elastomeric based sealant products. Work undertaken by Lacasse [5] has shown that fatigue behavior of polyurethanes in the laboratory can likewise be extrapolated to provide an indication of the performance potential of these sealants in actual building joints. It was also demonstrated that an analysis of the results of this dynamic fatigue test for these polyurethane products allowed for parallels to be drawn between the fatigue behavior of sealants and that of other elastomers based on the work of Bartenev et al.[7]. The Bartenev group showed that rubber fatigue occurring in elastomeric materials, such as unfilled SKS-30A rubber, can be characterized by the following relation:

$$N\varepsilon^{m} = C \qquad (1)$$

In this equation, ε is the predetermined maximum deformation of the sample (i.e., the amplitude of movement of the sealant specimen), N is the number of cycles to failure, and both m and C are constants related to the intrinsic fatigue properties of the material. Of these two constants, neither are dependent on N, and m is neither dependent on the temperature nor the method of testing.

This relation forms the basis for the fatigue analysis of construction joint sealants in the present investigation. Thus, owing to their elastomeric composition, the failure data from sealants evaluated in this study should produce similar results when plotted according to the relationship described in equation (1). Hence, a plot of log cycles to failure versus log strain for elastomeric materials should give a straight line with a slope of -m and an intercept of log C. If log N-log ε plots can be used to characterize sealant fatigue behavior, then a quantitative method exists for comparing the fatigue behavior of different sealants as well as for estimating the lifetime of sealed joints. Furthermore, this method potentially provides the basis for accelerating the test.

Two high performance silicone sealants were tested using the Karpati method and the results compared with unpublished work from 1986 in which three other types of silicone based sealants were similarly investigated [4]. All five sealants analyzed were fitted to log N-log ε plots in an attempt to establish a relationship between sealant fatigue behavior and the relationship discussed above. This work forms part of a study on evaluating cyclic fatigue as a means of assessing the performance of construction joint sealants, and focuses in this initial series, on evaluating the fatigue behaviour of silicone based products. The behaviour of polyurethane and polysulfide sealants subjected to cyclic fatigue testing will be reviewed in subsequent publications.

EXPERIMENTAL

Materials

The five silicone sealants, designated KS1, KS2;(KS2P), KS3, S01 and S02, all had similar application characteristics. The letter K in the designation refers to specimens evaluated in a study undertaken by Karpati and is presented here for comparitive purposes; the letter P indicates specimens that have been prepared with primer applied to the substrate. The postscript S refers to sealants evaluated in the present study. These products were all one-part moisture cured, gun-grade, non-sagging, commercial brands of silicone sealant having a gray color.

Procedure

The experiment was broken into three phases based upon ranges of amplitude deformations. The lower amplitudes of movement, ±10% to ±45%, were initiated in the first phase since failure time of these specimens was expected to be much longer. The second phase included amplitudes ranging from ±20% to ±60% in order to provide some overlap with specimens evaluated in phase one as well as to cover slightly higher amplitudes of movement. In the third phase, the sealant products were cycled from ±45% to ±80%.

For each phase of the experiment, 108 pairs of aluminum substrates with dimensions of 12.7 X 25.4 X 76.2 mm were first mechanically cleaned and then solvent wiped with xylene, and 36 pairs of these substrates were primed according to manufacturer's instructions. The silicones, S01 and S02, were applied to the non-primed substrates and the molds were lined with wax paper to prevent adhesion. In all cases, polyurethane spacers were used to maintain the size of all sealant beads at 12.7 X 12.7 X 50.8 mm.

All specimens tested in this set of experiments were cured for approximately 48 days at 21±2°C and 50±4% R.H. and were then attached to hand-operated vises [6] as shown in Figure 1. The silicone sealants tested by Karpati (K1,K2, K2P, K3) were all cured for one day prior to being mounted on the vises.

In all phases of the study, sealant specimens were subjected to repeated cycles of extension and compression at a rate of 7.5 cycles per week [3]. The use of a cycling rate of 7.5 cycles per week allows for alternating nights as well as alternating weekends of extension and compression. The cycling schedule is shown in Figure 2.

Figure 1. Hand operated vise used to subject sealant specimen to cyclic movement

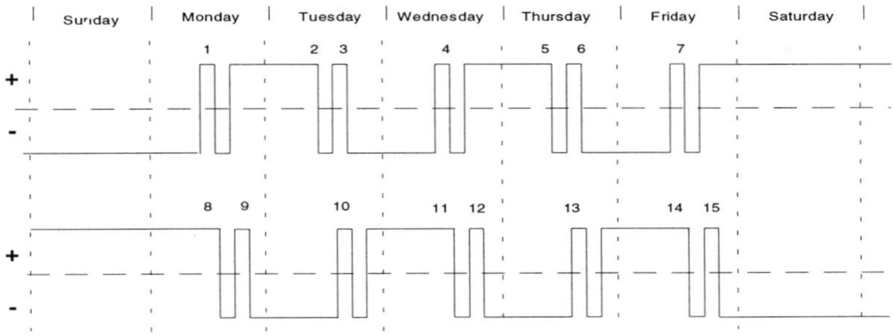

Figure 2. Schematic diagram of cycling schedule.

Within each test phase of this study, and for all amplitudes of joint movement, four specimens of each of the two sealants were evaluated (Refer to Table 1 for details). Cyclic movements of extension and compression were imposed on the samples by manually adjusting the vises to give amplitudes of movement ranging from ±10% to ±80% of original joint width, with amplitude increasing in 5% increments. Full extension was achieved to the desired strain amplitude within 5 to 20 seconds of the beginning of an amplitude adjustment.

TABLE 1 -- Number of specimens cycled at each amplitude in each phase

Amplitude of Movement (±%)	Number of Specimens at each Amplitude					
	Phase I		*Phase II*		*Phase III*	
	S01	S02	S01	S02	S01	S02
10	4	4	...	...	...	...
15	4	4	...	...	...	...
20	4	4	4	4	...	...
25	4	4	4	4	...	...
30	4	4	4	4	...	...
35	4	4	4	4	...	...
40	4	4	4	4	...	...
45	4	4	4	4	4	4
50	...	...	4	4	4	4
55	...	...	4	4	4	4
60	...	...	4	4	4	4
65	...	...	...	...	4	4
70	...	...	...	...	4	4
75	...	...	...	...	4	4
80	...	...	...	...	8	8

Almost all samples were started in compression (denoted by negative percentage) so that in extension (denoted by positive percentage) all samples had been through an integer number of cycles. The only exception occurred in phase III when eight of each of S01 and S02 were cycled at ±80%, with four of each being started in compression and four in extension. This was a check to see if the initial cycle position of the specimen affected the number of cycles to failure. The failure data for the samples at ±80% extension/compression, which were initially extended, were not included in the analysis since this change produced different results. Compression of samples to very small joint widths, as in the case of ±80% samples, often resulted in cutting of the sealant along the edge of the substrate. Hence, cracks and tears began a half-cycle earlier for those samples which were first compressed. Therefore, in order to use the same basis of evaluation for the different sealant specimens, only those samples compressed in the first cycle were included in the final analysis.

The manipulation of the phase I samples began on April 5, 1993. The movements imposed on the remainder of the specimens in this phase ranged from ±10% to ±45% of original joint width. Cycling of the phase II samples began on May 17, 1993, and for this set of specimens, amplitudes of imposed movement ranged from ±20% to ±60% of original width. In phase III of the experiment, initiated on July 20, 1993, the silicone sealants were cycled from ±45% to ±80%.

After each cycle, samples were examined for failure while still in extension, just before they were compressed. Sealant failure was visually identified as it progressed through essentially three stages, as described below:

Mode A failure:	initial cracking/peeling at corners of sealant bead.
Mode B failure:	rupture/peeling at the corners or along the edges of the sealant/substrate interface extending to nearly half the width of the specimen.
Mode C failure:	complete rupture of sealant bead.

The number of cycles taken to reach each mode of failure was recorded and samples were removed from testing after mode C failure occurred. As discussed by Karpati [3], it is mode C failure which gives useful results when extrapolating for long periods of time since logarithmic plots of both mode A and mode B failure do not necessarily define a linear relationship, but rather a broad band of data which reduces the accuracy of the results. For this reason, it was the number of cycles to mode C failure that was used in the analysis of the results. Photographs of the samples were taken while the sealant beads were still extended, just before compression, and were used as a visual record of failure progression.

RESULTS AND DISCUSSION

The results of fatigue tests on sealant products evaluated in this study (SO1, SO2) are compared to data from unpublished results obtained by Karpati based on an evaluation of three different silicone based sealants (KS1, KS2, KS3). Depicted in Figure 3 is a representative progression of failure for an S01 sample cycled at ±60% extension/compression at six cycles (Mode A failure), eight cycles (Mode B failure), and twelve cycles (Mode C failure). Not only does a tear result in this case, but the S01 sample also undergoes plastic deformation as evidenced by the inability of the sample bead to return to its original undeformed shape at the end of each cycle.

Figure 4 is a series of photographs showing the progressive failure of an S02 sealant specimen undergoing extension/compression of ±60% joint width. At two cycles, the specimen is still at mode A failure (tiny cracks at back of sample), but by the fourth cycle, a large tear has developed (mode B failure) and this quickly leads to complete failure of the bead at five cycles. This cohesive failure is typical of the failure mode in S02 specimens.

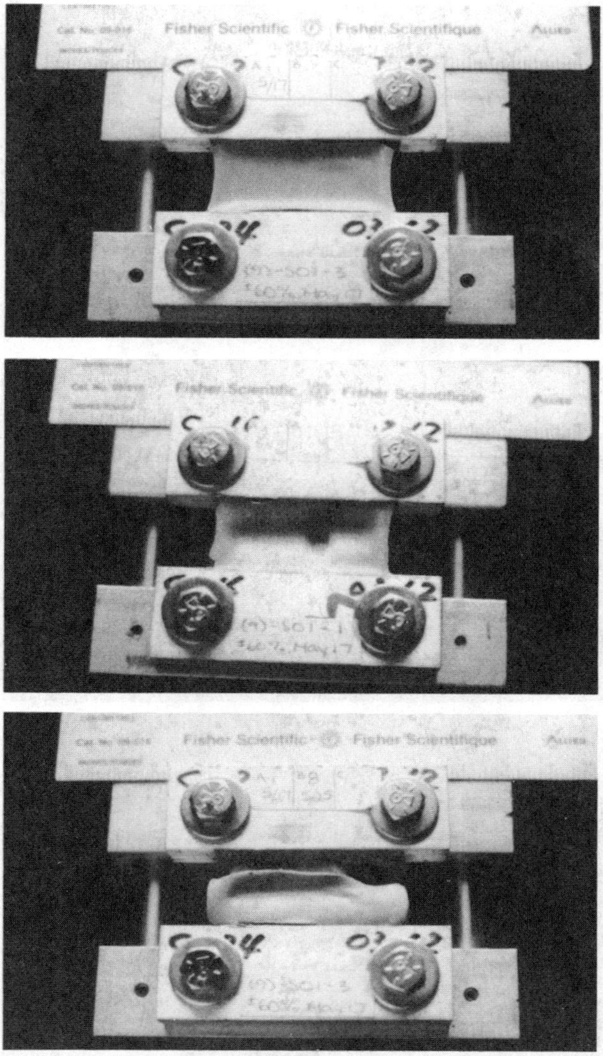

Figure 3. Series of photographs depicting various modes of failure for SO1 type sealant specimens; cyclic movements of ±60% of nominal 12.7 mm (0.5 in.) joint width. Top (mode A failure): tearing at the top left & right and loss of bond along bottom edge of sealant bead after 1 cycle; Middle: formation of hole at center of sealant bead indicates imminent rupture & onset of mode B failure after 7 cycles; Bottom (mode C failure): complete rupture of bead after 11 cycles.

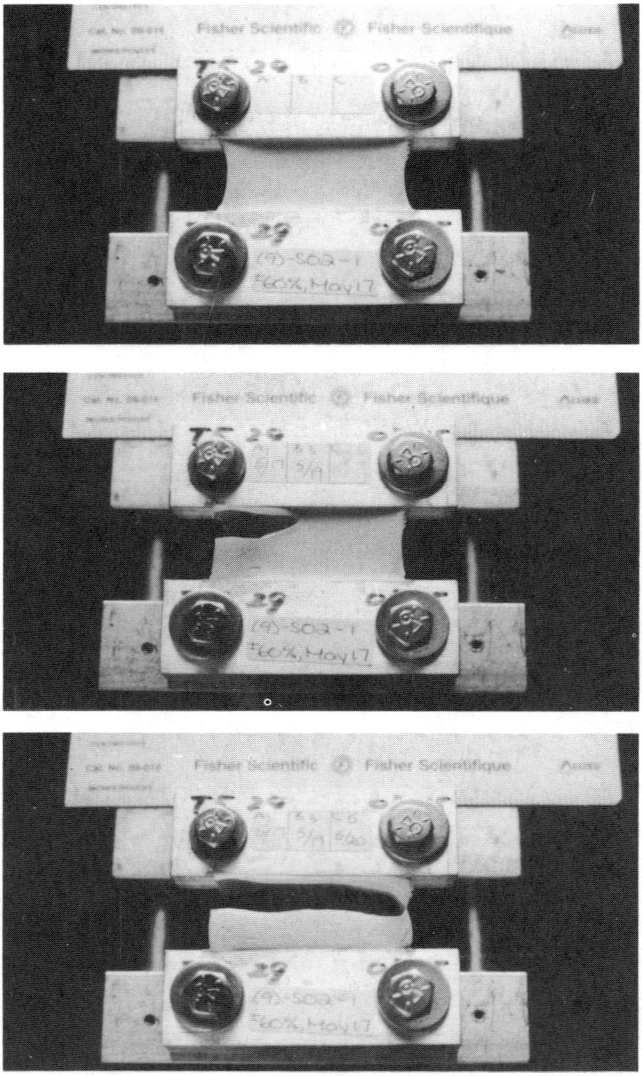

Figure 4. Series of photographs depicting various modes of failure for SO2 type sealant specimens subjected to cyclic movements of ±60% of nominal 12.7 mm (0.5 in.) joint width. *Top*: tearing at the top left of sealant bead after 1 complete cycle indicative of mode A failure; *Middle*: initial rupture of sealant bead after 3 cycles shows mode B failure; *Bottom*: complete rupture of bead after 5 cycles illustrates mode C failure.

While photographs can give a good qualitative picture of the failure of these sealants, mathematical analysis is required if a more quantitative perspective is desired. For this reason, the number of cycles to mode C failure (N) was used to calculate log N and the amplitude of movement of the specimens was used to determine the strain (ε) on the samples. The values of log N at each strain for each sealant were averaged and log N versus log ε was plotted for all specimens.

For all sealants, regression of log N-log ε plots (figures 5 -10) did indicate a linear relationship as described by Bartenev in equation 1. However, the strength of this relationship based on 95% confidence limits varied for the various sealants and appeared to be strongest for sealants having the larger sample size tested to the greatest number of cycles, as in the case of S01 (Fig. 5) and S02 (Fig. 6). As displayed in Figure 7, the data for silicone KS1 shows a weaker linear log N-log ε relationship than silicones KS2 (Fig. 8), KS2P (Fig. 9), and KS3 (Fig. 10) which showed good agreement with their respective regression lines, having most of their data points falling within the 95% confidence bounds of the regressions.

Since plots of log N-log ε for these sealants show the expected linear behavior, the constants m and C, as determined by regression, should be useful in making joint lifetime predictions. Perhaps the best way to examine the relative behavior of these sealants is then to examine the expected lifetimes calculated using the constants from Table 2.

TABLE 2 -- <u>Value of constants from equation (1), as determined by regression</u>.

Sealant	m	C
KS1	1.35	1.40
KS2	2.54	0.42
KS2P	1.93	1.21
KS3	1.13	1.57
S01	4.30	0.651
S02	4.26	0.671

From values of m and C, potential sealant lifetimes (in cycles to mode C failure) at different amplitudes of movement can be calculated based on the number of cycles to failure and the rate of cyclic movement (7.5 cycles /week). An estimate of the useful lifetime of joints sealed by the five sealants under examination are given in Table 3.

In general, a smaller value of m indicates a comparatively reduced estimate of lifetime when evaluating sealants at a given movement amplitude. Smaller values of m also provide reduced sensitivity in

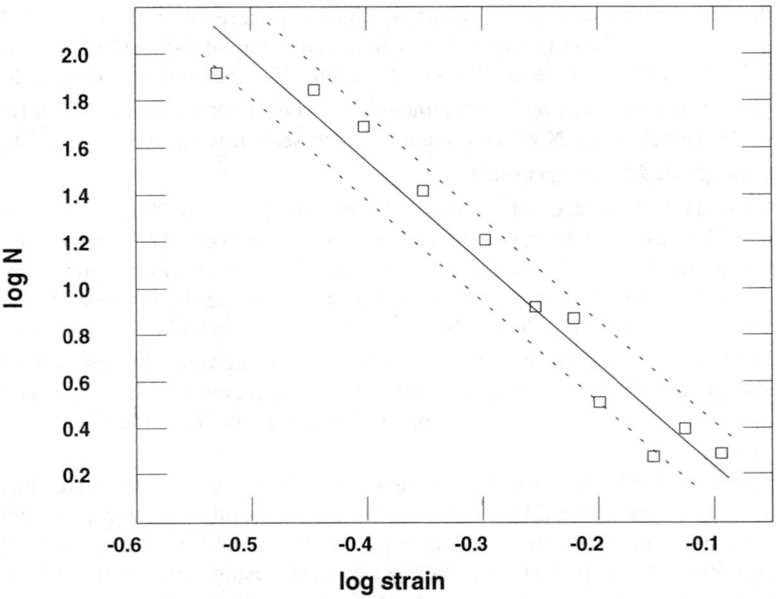

Figure 5. Linear regression (with 95% confidence limits) of log N - log strain plot of S01 data.

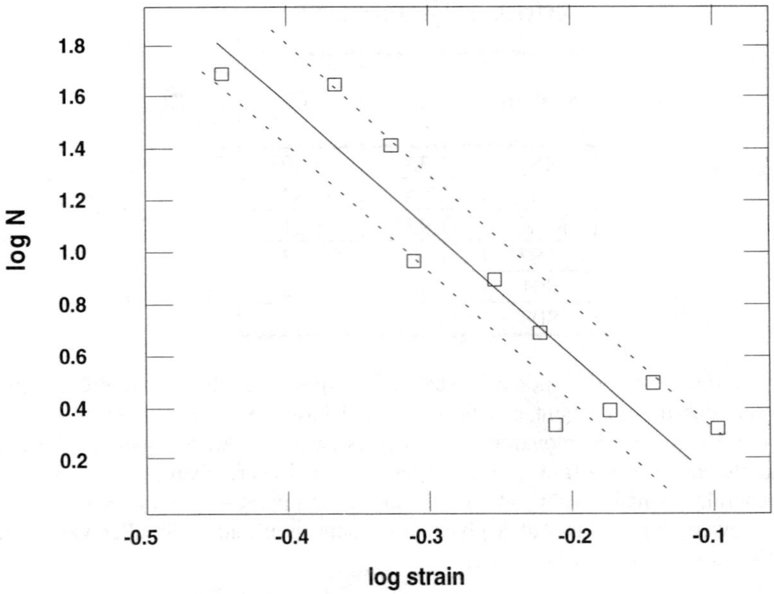

Figure 6. Linear regression (with 95% confidence limits) of log N - log strain plot of S02 data.

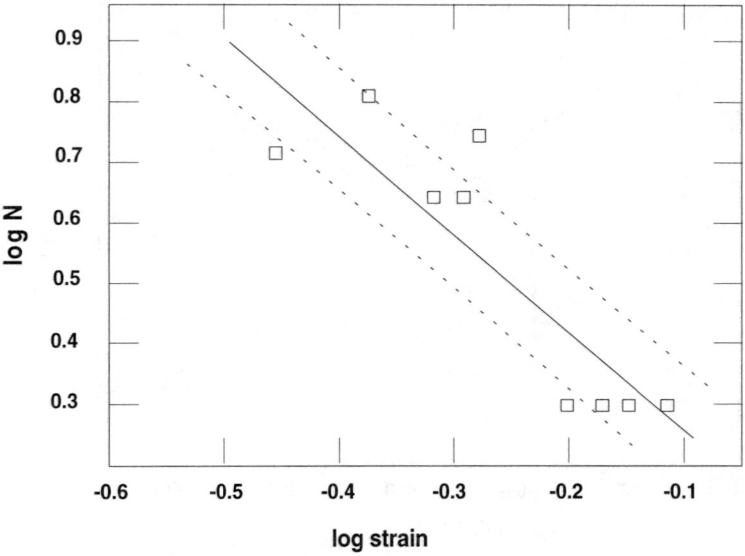

Figure 7. Linear regression (with 95% confidence limits) of log N - log strain plot of KS1 data.

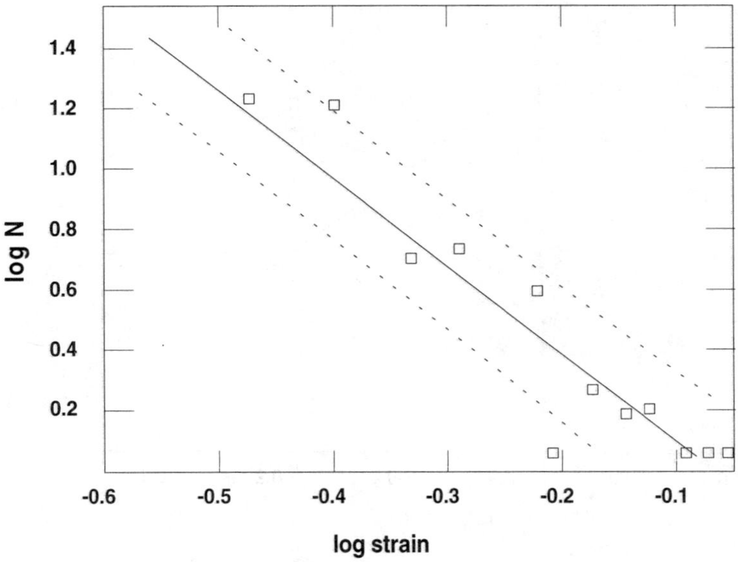

Figure 8. Linear regression (with 95% confidence limits) of log N - log strain plot of KS2 data.

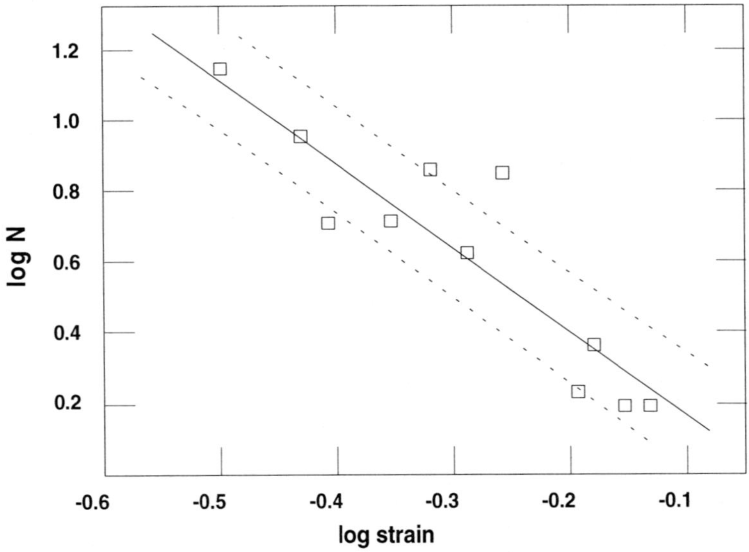

Figure 9. Linear regression (with 95% confidence limits) of log N - log strain plot of KS2P data.

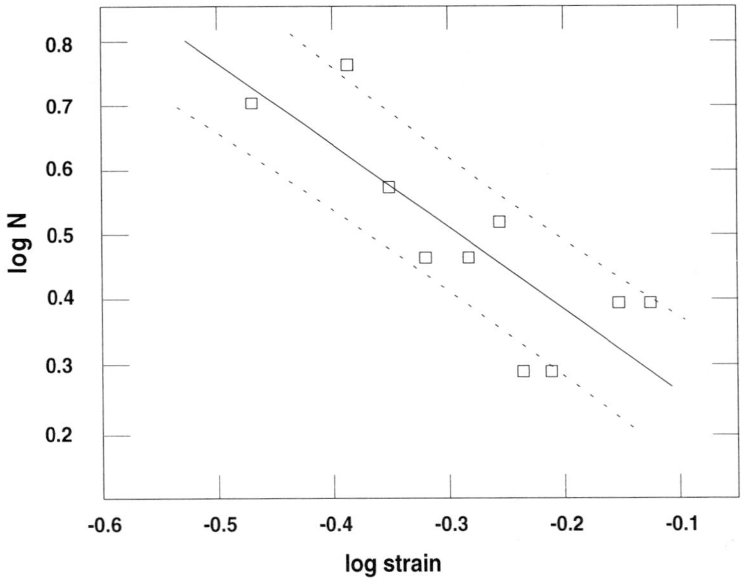

Figure 10. Linear regression (with 95% confidence limits) of log N - log strain plot of KS3 data.

TABLE 3 -- Estimated joint lifetimes for all sealants analyzed

Sealant	Joint lifetime in years for given amplitudes of cyclic movement				
	±15%	*±20%*	*±25%*	*±30%*	*±35%*
KS1	0.27	0.24	0.22	0.21	0.20
KS2	0.75	0.48	0.34	0.26	0.21
KS2P	0.68	0.52	0.43	0.36	0.31
KS3	0.20	0.19	0.18	0.18	0.18
S01	33.14	12.81	6.13	3.35	2.01
S02	31.58	12.36	5.97	3.29	1.99

estimated service-life for changes in amplitude of cyclic movement. This is especially noticeable in the case of KS3, where changing the amplitude from ±15% to ±35% has very little effect on the expected lifetime of 0.2 years. Evidently, the predicted service-life is also affected by the constant C. Since log N is shifted (up or down) by the amount log C, a larger value of C, and therefore of log C, will also increase the predicted lifetime of the joint.

The effect of the primer on sealant K2 is not entirely clear. While K2P does have a smaller value of m, and is therefore less sensitive to amplitude changes, both sealants display fairly similar behavior in the range of joint widths which are typically used (i.e. ±15% to ±30%, refer to Table 3). further testing would be required to elucidate the significance of using primer for this type of sealant.

One potential difficulty with analyzing Karpati's data was that KS1, KS2, KS2P, and KS3 were only cured for one day. Cycling began early on, before samples were fully cured, and thus the polymeric structure of the sealant material was not fully developed. While this may more accurately reflect actual sealant behavior in building joints, it could limit the usefulness of interpreting this particular set of data in the same way as that of a typical elastomer undergoing cyclic fatigue testing or of comparing it to more fully cured samples.

Other factors that should be considered are the very limited number of samples evaluated in the Karpati study and the fact that these tests were carried out until failure or 25 cycles, whichever came first. It is useful to include the results here as they demonstrate the significance of the relationship and the characteristic changes in values of m and C for less rigorously compiled data. Another aspect which should be considered is that the lack of thorough testing in the Karpati study places accent on premature failures, which tends to limit the usefulness of these results since they do not necessarily accurately reflect normal sealant performance. Yet another factor is that Karpati did not clearly indicate which failures were purely adhesive. Hence, the data points are potentially a mix of cohesive and adhesive failures necessarily skewing results to favour a reduction in the estimated service-life of the sealant. It is probably for the above reasons that, while a linear relationship does exist between log N and log ε, the joint lifetime predictions for the silicones tested by Karpati are apparently low. Had tests been carried out until all samples failed, rather than just recording the samples which failed early on, it is surmised that the estimated lifetimes would yield life-time estimates more closely aligned with those estimated for S01 and S02.

For sealants S01 (Fig. 5) and S02 (Fig. 6), the testing of 108 of each of these sealants ensured that enough samples had failed to adequately describe failure trends. The strength of the linear log N-log ε relationship for these sealants is a clear indication of the potential usefulness of interpreting fatigue results for sealants in this manner. The estimated lifetimes obtained by use of the regression data are reasonable enough to assume that this is a valid method of sealant performance evaluation.

The ranking of silicone products against each other is an issue difficult to address owing to the number of unknown factors in the experiments carried out by Karpati [4]. However, Figure 11 show how the log N-log strain plots of the sealants compare against each other, and graphically reflect the behavior reported in Table 3. The similarity in the fatigue behavior of S01 and S02 is evident in this figure as their regression lines are coincident, whereas the shallow slopes of the sealants KS1, KS2, KS2P, and KS3 indicate their inability to survive long periods of cyclic deformation.

Typically, silicone sealants are said to have movement capabilities of 25% (i.e. amplitudes of joint width movement of ±25%). Ranking the sealants according to their ability to meet their expected performance level gives the results displayed in Table 4. It is interesting to note that the silicones S01 and S02 do not perform as well at the higher amplitudes that they are propounded as being able to withstand.

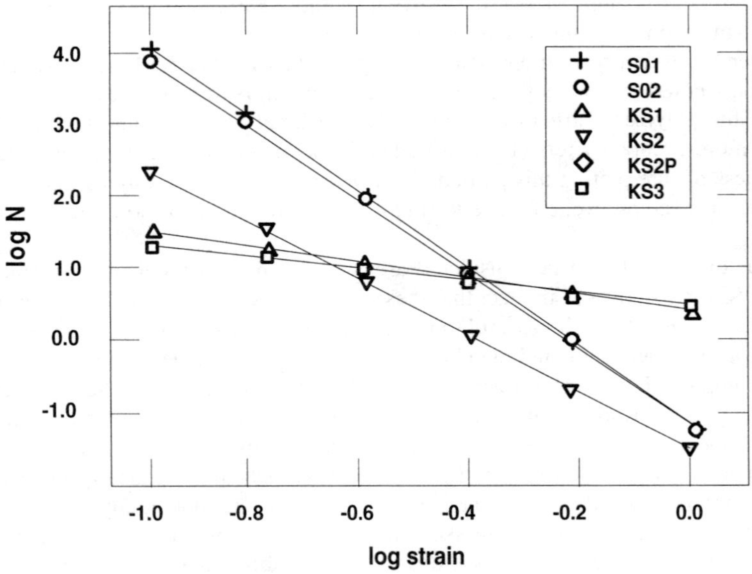

Figure 11. Log N - log strain regression lines for all silicone sealants.

TABLE 4 -- Ranking of sealants by their ability to meet the expected movement rating of their class

Sealant	Movement Rating	Lifetime (years)
S02	50%	0.622
S01	50%	0.621
KS2P	50%	0.223
KS1	50%	0.173
KS3	50%	0.167
KS2	50%	0.118

CONCLUSIONS

It has been shown that sealant fatigue behavior can be analyzed using the same principles that Bartenev models elastomeric fatigue [9]. In this manner, fatigue behavior of different sealants can be compared quantitatively through joint lifetime predictions. However, while results shown here do demonstrate the potential usefulness that this method of analysis has for the study of sealant fatigue, there are precautions that should be taken to ensure useful results.

The first focuses on attempting to limit the damage that occurs to the sealant both during casting and, in the case of samples extended and compressed above ±50%, during compression when the edges of the substrates can cut the bead as it is displaced. Such cuts or flaws can promptly initiate crack growth which might not otherwise have occurred if the sealant bead had not been damaged. While this might reflect actual service conditions of a sealed joint, it would limit the usefulness of studying sealants as elastomers under more ideal conditions. Thorough testing is another factor which greatly improves the quality of the results. Quite obviously, the larger the number of samples tested, the more accurately the results will reflect those of a larger population of samples. Finally, testing until failure, and not just until a predetermined cut-off point, ensures that it is not simply premature failures that are included in the results.

Potential improvements to the test would include the testing of joints that have been cast in a variety of shapes such that sealant performance can be evaluated as a function of joint geometry. Hence using the same methods employed in this study, the dependence of joint lifetime on the "shape factor" could be determined by changing the joint depth and width ratios in specimens being evaluated.

Based on the results of this study, 48 days cure does appear to be sufficient time to allow the polymeric network of these elastomeric sealants to fully develop. If sealants are to be evaluated based on the analytical precepts outlined previously, cycling should be initiated after the samples are fully cured. A study to compare the fatigue behavior of both partially and fully cured sealant specimens would help clarify the significance of partial cure on the fatigue life. This would be a further contribution to developing an accelerated test method useful in predicting the long-term performance of joint sealant products based on a knowledge of field conditions.

REFERENCES

[1] Karpati, K.K., "*Weathering of Silicone Sealant on a Strain-cycling Exposure Rack*", Adhesives Age, Vol. **23**, No. 22, 1980, p.41-47.

[2] Karpati, K.K., "*Investigation of the Factors Influencing the Outdoor Performance of Two-part Polysulphide Sealants*", Journal of Coatings Technology, Vol. **56**, No. 719, 1984, p.57-32.

[3] Karpati, K.K., "*Laboratory Fatigue Test of a Two-part Polysulfide Sealant Correlated to Outdoor Performance*", Durability of Building Materials, Vol. **5**, No.1, p.35-51, 1987.

[4] Karpati, K.K., Solvason, K.R., Sereda, P.J., "*Weathering Rack for Sealants*", Journal of Coatings Technology, Vol. **49**, No.626, 1977, p.44-47.

[5] Lacasse, M.A., Functional Properties of Polyurethane Based Sealants Blended with Polymeric Modifiers, Ph.D. Thesis, Centre for Building Studies, Concordia University, Montreal, 1991.

[6] Karpati, K.K., "*Device for weathering sealants undergoing cyclic movements*", Journal of Coatings Technology, Vol. **50**, No. 641, 1978, p.27-30.

[7] Bartenev, G.M. and Zuyev, Yu. S., Strength and Failure of Visco-Elastic Materials, Pergamon Press Ltd., 1968.

John Beech* and John Beasley*

EFFECTS OF NATURAL AND ARTIFICIAL WEATHERING ON BUILDING SEALANTS

REFERENCE: Beech, J., and Beasley, J., **"Effects of Natural and Artificial Weathering on Building Sealants,"** Science and Technology of Building Seals, Sealants, Glazing and Waterproofing: Fourth Volume, ASTM STP 1243, David H. Nicastro, Ed., American Society for Testing and Materials, Philadelphia, 1995.

ABSTRACT: A variety of commercial sealant products have been studied at BRE over the past 20 years, during which changes in key properties have been measured in samples exposed to artificial laboratory ageing and weathering, and to natural weather in England and in more severe climates abroad. Most recently the results of a detailed study of cure and durability of sealant products have been reported at ASTM Symposia in 1992 and 1993.

The results of this laboratory study have been compared with data obtained from samples exposed to natural weather in a temperate climate in Southern England and a middle eastern climate for 6½ years and in a tropical climate for 4 years. Highly significant correlations have been found between modulus changes after exposure for 40 and 52 weeks in an accelerated weathering apparatus, and the effects of weather in all three climates. The implications for the prediction of the durability of sealants from laboratory test data are discussed, and recommendations made for suitable test regimes in relation to the climates to which the sealants were exposed.

KEYWORDS: sealants, durability, weather, accelerated ageing

INTRODUCTION

BRE began a two-year experimental study of the cure and durability of building sealants in 1990, in which the tensile properties of suitable samples were measured after known periods of cure and subsequent exposure to natural weather and to various types of simulated weathering in the laboratory.

The results of the first stage of this study, using freshly cured samples, were reported at the 2nd ASTM Sealants Symposium[1]. This was followed by a paper reporting the second stage, using samples cured for 12 months, the following year[2].

In both stages of the experimental study it proved possible to demonstrate some highly significant correlations between properties (such as secant modulus and extensibility) when these were measured after 12 months exposure to natural weather in Southern England, and after artificial weathering using potential degradation factors such as elevated temperature, ultra-violet radiation and moisture.

These results, based on property changes of a range of high-

* Building Research Establishment, Garston, Watford, Hertfordshire, WD2 7JR, United Kingdom.

performance elastomeric sealants (polysulphides, silicones, and polyurethanes), provide some encouragement that artificial weathering may eventually allow reliable predictions of the long-term performance of sealants in their service environments.

However, one year exposure to temperate natural weather is unlikely to produce physical and chemical changes in these sealants which are typical of those occurring during longer periods, such as those a sealant must withstand in a building joint (typically for 20 years or more). It is therefore desirable to establish a relationship between property changes during laboratory weather simulations, and those after longer periods of exposure to weather, both in temperate and in more severe climates.

BRE has exposed high-performance sealants, available on the UK market, to weather in England and in the Middle-East (Dubai) and West Africa (Sierra Leone), over the past 15 years.[3,4] The data appertaining to the tensile properties of samples retrieved from these weathering sites have been examined, with a view to detailed comparison between relevant experimental data and the results of the recently reported durability study.

It was found that the results from sealant specimens exposed for between 9 and 15 years were not strictly comparable with the recent data, since although some of the products were ostensibly the same in the separate studies, in fact major changes in formulation had taken place in many of the sealants marketed during this period. It was concluded that this would invalidate any attempted correlations between the respective data for property changes in the sealants.

Data are however available for shorter periods of exposure to weather, in England and in the overseas sites, which were obtained using samples prepared from essentially the same sealant products as those used in the recent laboratory studies. Accordingly, a detailed analysis of the effects of exposure to weather on tensile properties has been made for samples exposed in Southern England (Garston, Hertfordshire and Prices Risborough, Buckinghamshire) and in Dubai for 4 and 6½ years; and Sierra Leone for 4 years.

This paper presents the results of comparisons between data from laboratory weathering and longer term natural weathering in temperate and more severe climates.

EXPERIMENTAL

Preparation of specimens

The test specimens each comprised a 12 x 12 50 mm bead of sealant between the large surfaces of cement mortar and aluminium adherents. The cement mortar pieces were 12 x 50 x 8 mm in dimension and were prepared by the method described in BS 4254[6]. Aluminum test pieces were 12 x 50 x 10 mm, with a milled finish.

Five replicate specimens were prepared of each sealant for each of the test exposures, using the manufacturers' recommended primers for the two substrates. The specimens were stored at 23°C and 50% RH (at 27°C and 95 % RH in the case of P13) to allow initial cure to take place, drawing the spacer bars from the sealant after a few days in the case of the moisture curing materials.

Materials

Nine commercial sealants were used in the tests. The coding below is used in the paper.

P13	Black	One part polysulphide conforming to BS 5215: 1983[5]
2P10	Grey	Two part polysulphides conforming to BS 4254: 1986[6]
2P15	Black	
S6	Grey	Low modulus silicone sealants
S12	Brown	conforming to BS 5889: 1989[7]
S13	Black	High modulus silicone sealants
S11	White	conforming to BS 5889 : 1989[7]
		(S11 and S13 differ only in colour)
U1	Grey	One part polyurethane sealants
U6	White	
U3	Grey	Multi part epoxy polyurethane

Test Exposure

The specimens were allocated randomly to the exposures listed below, for each set of specimens made using one of the nine products. As mentioned previously a second set of each product was set aside for 12 months storage at standard conditions, for use in the second phase of the study.

The test exposures used were:

Control (CN) : stored at standard conditions of 23°C ± 2°C and 50 ± 5% RH.

Heat (HT) : storage in an oven at 70°C ± 2°C.

Ultra Violet Light with Condensation (UC) : artificial weathering in a QUV apparatus with cycles of 4 hours of UV(B) radiation at 70°C followed by 4 hours condensation at 40°C.

Natural Weather (NW)

Samples were exposed facing south on racks inclined at 45°, on four weathering sites:

United Kingdom : Prior to 1988 at Princes Risborough and subsequently at Garston, a temperate climate in southern England.

Dubai., Middle East : a warm, dry climate, close to the sea, with long periods of solar radiation by day and a rapid fall in temperature by night for some parts of the year.

Sierra Leone, West Africa : hot humid Equatorial climate with little seasonal or diurnal temperature variation, high-intensity diffuse UV radiation.

The coding above is used throughout the paper.

Tensile Testing

Five specimens of each sealant were removed from each of the five test environments after the following duration of exposure: 2, 4, 8, 16, 28, 40 and 52 weeks. They were conditioned for 24 hours at standard conditions before being placed in a tensile test machine and pulled to destruction at 5mm/min. Force data were recorded at 0.1mm increments of strain and stored on disc for subsequent analysis.

A set of five control samples of each sealant was also tested in this manner after the initial storage period of 28 days under standard conditions.

From the tensile test data the following parameters were obtained for each specimen: force (N) at 15mm joint width, which was taken as equivalent to the secant modulus at 25% extension; the maximum force before break was obtained; and the elongation at this force.

Specimens retrieved from the overseas exposure sites were stored in the laboratory at 23°C and 50%RH for at least 7 days before extension at 5 mm/min in a tensile test machine to a width of 15mm.

RESULTS & DISCUSSION

Comparison of Laboratory and Natural Weathering Data

In the two previous papers[1,2] reporting the results of the two phases of the laboratory studies of cure and durability, a number of significant correlations were found between the laboratory data and those for 12 months exposure to weather in Southern England. In particular, highly significant correlations were obtained, using freshly cured specimens, for changes in modulus between 8 weeks exposure to artificial weathering and 52 weeks natural weather.

In the second phase of the work, using specimens which had been stored for 12 months under standard laboratory conditions, cure effects were relatively small, and fewer significant correlations between the effects of short-term weather and the laboratory exposures were found. Nevertheless, two highly-significant correlations (i.e., at the 0.1 per cent level) were obtained: between extensibility data for 8 weeks heating at 70°C and weather; and between modulus values after 8 weeks exposure to QUV conditions and weather.

It was concluded that both artificial weathering using a cycle of UV(B) radiation and condensation, and exposure to the individual weather factors, such as heat and ultra violet radiation, had potential for prediction of the changes in sealant properties which occur during exposure to weather in a temperate climate. However it was recognised that a more rigorous test of their predictive ability would arise when attempting to relate these laboratory test data to those obtained after longer durations of weather exposure, more typical of the life-times of sealants used in building joints.

Data obtained from tensile measurements on specimens retrieved from exposure sites in England, West Africa, and the Middle East over the past 15 years were accordingly examined and compared with the data from the more recent laboratory study. It was concluded that the materials from these exposure sites which had been exposed for more than 7 years were in many cases sufficiently different from the later materials, because of changes in the formulation of ostensibly the same products, to render invalid any attempt to correlate the respective data.

Analysis of the data was therefore confined to data from the sites in England and Dubai for exposure durations up to 6½ years; and to data from Sierra Leone for a 4 year exposure period. In the cases of all the nine sealant products used, the material used in the two separate experiments were essentially the same, although the samples were prepared at different times and were not therefore drawn from the same population. In the case of one silicone product, designated S11 and S13 , a different colour was used for the weather exposure than for the laboratory data.

Correlation of Modulus Change Data

It was necessary to confine data analysis to changes in modulus relative to the initial mean control values since specimens removed from the exposure sites were not tested to destruction, and therefore no data for extensibility were available. The data examined by linear regression were obtained from 9 sealant products, after 6 different exposures to natural weather, and after exposure to the three laboratory tests for 2, 4, 8, 16, 28, 40 and 52 weeks.

Table 1 gives a summary of the significant correlations obtained where well-cured specimens were used in the laboratory measurements. Those obtained using data from laboratory tests on freshly-cured specimens are listed in Table 2.

Laboratory Test		Correlation with weathering		
Condition	Duration	Location	All (9) Sealants	Silicones and Polyurethanes
UV/Heat	8 weeks	Dubai - 2 years	5%*	-
UV/Cond	16 weeks	UK - 6½ years	5%	-
UV/Cond	28 weeks	UK - 6½ year	5%	-
UV/Cond	40 weeks	UK - 6½ years	1%	-
UV/Cond	40 weeks	Dubai - 2 years	5%	-
UV/Cond	40 weeks	Dubai - 4 years	1%	0.1%
UV/Cond	40 weeks	Dubai - 6½ years	5%	0.1%
UV/Cond	40 weeks	S.Leone - 4 years	5%	-
UV/Cond	52 weeks	UK - 2 years	1%	-
UV/Cond	52 weeks	UK - 6½ years	0.1%	-
UV/Cond	52 weeks	Dubai - 2 years	5%	-
UV/Cond	52 weeks	Dubai - 4 years	5%	0.1%
UV/Cond	52 weeks	Dubai - 6½ years	5%	0.1%
Heat only	8 weeks	UK - 6½ years	5%*	-
Heat only	16 weeks	UK - 2 years	5%*	-
Heat only	16 weeks	UK - 6½ years	5%*	-

* Negative correlations

Table 1: Summary of significant correlations of % modulus change between natural weathering and well cured laboratory test specimens.

It is immediately apparent, having regard to all the correlations of at least 5 per cent significance level, that the laboratory data obtained with freshly-cured specimens tend to show positive correlations with shorter periods of weather exposure; whereas those for the well-cured laboratory specimens give correlations (generally of higher significance) with the data for longer periods of weather exposure in England and for the hotter climates. This is consistent with the findings of our earlier paper reporting the laboratory study on freshly-prepared specimens[1]. It is explicable in terms of the greater influence of cure phenomena on modulus changes in short-duration exposures to a moderate climate.

Before placing specimens on the exposure sites in Dubai and Sierra Leone, it was necessary to store them for some months in the laboratory until a suitable opportunity to visit these sites occurred. Typically these specimens were then comparable in their state-of-cure with the laboratory-tested specimens which were stored for 12 months to facilitate cure. The possible exception is the one-component polysulphide, P13, of which control samples showed significant increases in modulus even 12 months after preparation. It can also be reported that specimens of this type of sealant were found to be incompletely cured in the centre of the material, when removed from the Dubai site after one year.

Laboratory Test		Natural Weathering		Significance Level
Conditions	Duration	Location	Duration	
UV/Cond	2 weeks	UK	1 year	5%
UV/Cond	4 weeks	UK	1 year	5%
UV/Cond	4 weeks	UK	2 years	5%
UV/Cond	8 weeks	UK	1 year	5%
UV/Cond	8 weeks	UK	2 years	5%
Heat only	2 weeks	UK	1 year	5%
Heat only	4 weeks	UK	1 year	5%
Heat only	4 weeks	UK	2 years	5%
Heat only	28 weeks	Dubai	4 years	5%*
Heat only	28 weeks	Dubai	6½ years	5%*
Heat only	28 weeks	S Leone	4 years	1%*
Heat only	40 weeks	S Leone	4 years	5%*

* Negative correlations

Table 2: Summary of significant correlations of % modulus change between natural weathering and freshly cured laboratory test specimens.

Considering first the correlations found between the effect on modulus of heat alone and the various weather exposures in virtually all cases correlations were at the lowest level of significance (5%), in view of large deviations in the mean data values from the linear regression line, and the absence of correlations for longer periods of weather exposure, this test (though simple to perform) appears to have little merit as a predictive laboratory technique. It is notable that the only correlation of higher significance (viz 28 weeks of heat vs Sierra Leone exposure for 4 years, at 1% level) is, like others tabulated in both Tables 1 and 2, a negative correlation.

In practical terms this shows that sealants exhibiting large increases in modulus during 8 to 40 weeks at 70°C were most prone to show significant softening during long duration or severe weather exposures. P13 and U3 were typical of such sealants when freshly-prepared specimens were used for the laboratory exposures. These two materials, together with the two 2-part polysulphide products, showed similar behaviour when the laboratory specimens were well-cured. Although this provides a possible, simple means of identifying sealants which may be unsatisfactory for use in hot climates, it is not to be preferred to the more rigorous simulations of weather factors in combination.

The use of UV(B) radiation at 70°C appears to have little if any advantage over heat alone. A number of correlations of 5% significance are listed in Table 2 where comparison is made with short-duration weather exposure data for the UK.

The relationships which were found to be significant at 1% level or greater (in Table 1) are illustrated in Figures 1 to 5.

The data obtained using exposure in a "QUV" accelerated weathering machine, utilising UV(B) radiation from fluorescent tubes and condensation, offers the most promising means of prediction of the modulus changes for a range of elastomeric sealants, after exposure to longer periods of weather in England, and, to a lesser extent, in the more severe climates (Table 1).

It will be noted that the highly significant correlation (at the 0.1% level) between modulus changes after artificial weather for 52 weeks (Figure 1) and to temperate weather for 6½ years is consistent with the earlier findings of comparisons between natural and artificial weathering effects. The "acceleration factor" from the ratio of the durations in the laboratory and for natural weather is 6.5. This is the same ratio as was derived from the earlier correlation for UC and UK weather data - though it should be noted that the slopes of the regression lines are not identical in the two cases.

Nevertheless, this provides striking confirmation of a consistent relationship between the effects of artificial and of natural weathering, based upon the measurement of a "performance index", changes in which can be expected to have a direct relationship to the performance of sealants in service. However, it must be emphasised that all the data reported here were obtained from specimens which were exposed in an unstressed condition. It has not yet proved possible to establish reliably a consistent relationship between modulus changes in particular sealants and their ability to withstand repeated extension and compression while exposed to weather.

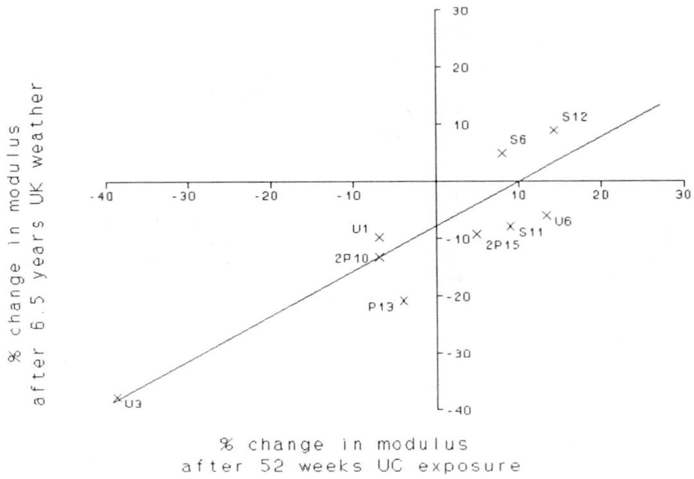

Figure 1 Correlation of 52 weeks UC data with 6.5 years UK weather.

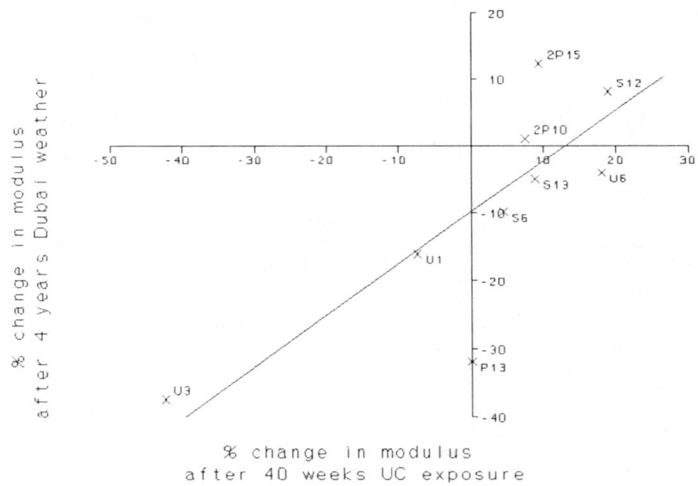

Figure 2 Correlation of 40 weeks UC data with 4 years Dubai weather.

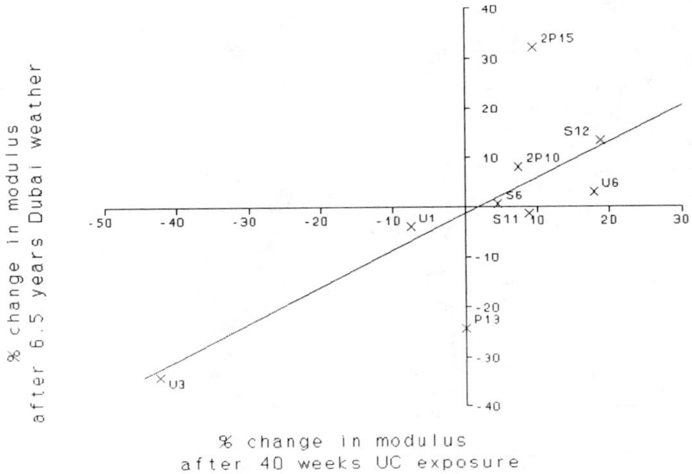

Figure 3 Correlation of 40 weeks UC data with 6.5 years Dubai weather.

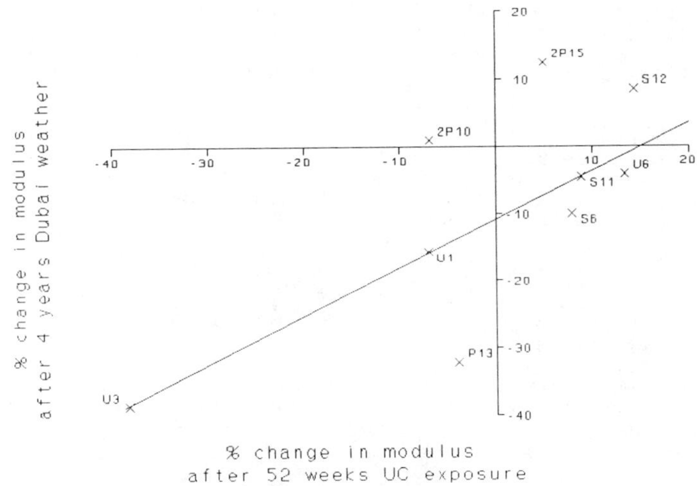

Figure 4 Correlation of 52 weeks UC data with 4 years Dubai weather.

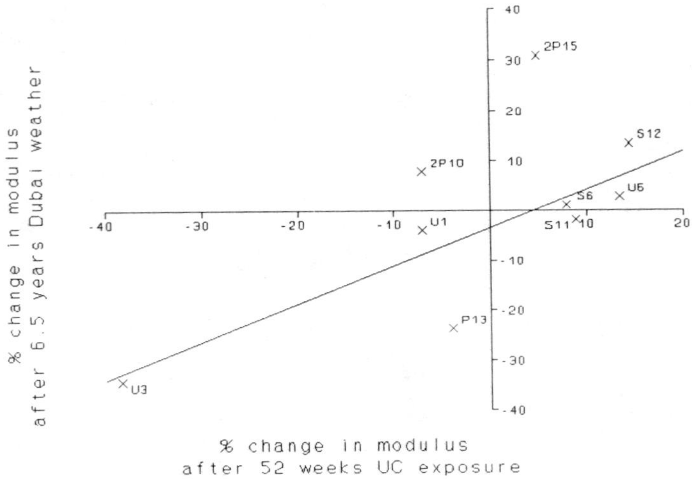

Figure 5 Correlation of 52 weeks UC data with 6.5 years Dubai weather.

The data for modulus changes in specimens retrieved from Dubai after 4 and 6½ years exposure are generally well-correlated with the effects of artificial weather for 40 and 52 weeks (Figures 2, 3, 4 and 5). It is apparent that in all these exposures, the 3 polysulphide materials deviate markedly from the general regression line. By eliminating these products from the regression calculation, correlations at the highest level of significance (0.1%) were obtained.

It is noteworthy that data from the Sierra Leone samples correlate less well with laboratory data : one correlation, between 40 weeks of artificial weather and 4 years exposure on the West Africa site, is listed in Table 1 (at the 5% level of significance). At first sight this is perhaps surprising, since the hot humid climate, with high intensity of diffuse UV, appears more similar to the simulated weather conditions in the artificial weathering apparatus than that of Dubai. In reality the Dubai climate, close to the sea and with large diurnal swings of temperature, providing conditions conducive to condensation at night, is probably well replicated in the weathering apparatus.

The linear regression lines for the data illustrated in Figures 1 to 5 may be represented as follows (only correlations significant at the 0.1% level are given): Modulus changes are expressed as percentages of the initial values.

(1) 52 weeks accelerated weathering (UC) vs 6½ years UK weather:

ΔM (UK) = 0.5 ΔM (UC) - 11.0
(for all 9 sealants)

The relationships in (2), (3), (4) and (5) below are valid for the silicone and Polyurethane products only (though similar linear regressions of lower significance were obtained when all 9 sealants were included - see Table 1).

(2) 40 weeks UK vs 4 years Dubai weather:

$\Delta M\ (D) = 0.65\ \Delta M(UC) - 10.6$

(3) 40 weeks UK vs 6½ years Dubai weather:

$\Delta M\ (D) = 0.68\ \Delta M\ (UC) - 4$

(4) 52 weeks UC vs 4 years Dubai weather:

$\Delta M\ (D) = 0.724\ \Delta M\ (UC) - 10.5$

(5) 52 weeks UC vs 6½ years Dubai weather:

$\Delta M\ (D) = 0.763\ \Delta M - 3.7$

Finally, it should be stressed that the derivation of precise relationships from these correlations does not permit the accurate calculation of modulus changes in laboratory tests. Even with correlations of high significance, deviations of specific data points from the regression line can introduce large errors. Furthermore, it must be borne in mind that the mean values used to derive these data each represent variability in the respective measured values of comparable magnitude to these deviations.

CONCLUSIONS

1. Highly significant correlations have been found between modulus changes after long periods of exposure in an accelerated weathering apparatus (40 and 52 weeks) and those measured on sealants specimens exposed to weather: for 6½ years in England, and for 4 and 6½ years in Dubai (the latter for silicone and polyurethane sealants only).

2. Other significant correlations are reported between data from laboratory simulations of weathering and aging, and the effects of exposure to weather in temperate and more severe climates, for the nine elastomeric sealants.

3. Significant negative correlations between modulus increases after heating at 70°C, and changes with exposure to weather in both temperate and more severe climates suggest the possibility of a simple aging test to identify sealants prone to poor performance in service.

4. The data for correlations between laboratory and natural weather exposures have confirmed the importance of using well-cured specimens for the prediction of long-term performance by laboratory simulations of weather.

5. Accelerated weathering based on UV(B) radiation from fluorescent light sources and condensation appears from the correlation data reported to be more appropriate to the temperate and middle-eastern climates than to the equatorial climate.

BIBLIOGRAPHY

(1) Beech, J C, and Beasley, J L. "Evaluation of Cure and durability of buildings sealants". Science and Technology of Building Seals, Sealants, Glazing and Waterproofing, Second Volume, ASTM STP 1200, J M Klosowski Ed., American Society for Testing and Materials, Philadelphia, 1992.

(2) Beech, J C, and Beasley, J L. "Further studies of Cure and durability of building sealants". Science and Technology of Building Seals and Sealant, Third Volume ASTM STP 1254, J Myers Ed., American Society for Testing and Materials, Philadelphia, 1993.

(3) Beech, J C, and Beasley, J L. "Assessment of the durability of building sealants". Proc. RILEM 1st International Congress "From Materials Science to Material Engineering", Paris, 1987. Vol 3, 1178-1185, Chapman & Hall, 1987.

(4) Beech, J C, and Mansfield, C. "Prediction of the durability of building sealants", Fourth Int. Conf. on Durability of Building Materials & Components, Singapore, 1987, Vol 1, 33-40, Pergamon Press, 1987.

(5) British Standard Specification BS 5215 : 1983. One part gun grade polysulphide based sealants. BSI. London, 1983.

(6) British Standard Specification BS 4254 : 1986. Two part polysulphide based sealants. BSI. London, 1986.

(7) British Standard Specification BS 5889 : 1989. One part gun grade silicone based sealants. BSI, London, 1989.

Products and Applications

Virginia K. O'Neil[1], Jerome M. Klosowski[1], Michael G. Altes[1] and Louise C. Bergman[1]

EFFECTS OF WEATHERPROOFING SEALANTS ON BUILDING AESTHETICS

REFERENCE: O'Neil, V. K., Klosowski, J. M., Altes, M. G. and Bergman, L. C., **"Effects of Weatherproofing Sealants on Building Aesthetics,"** Science and Technology of Building Seals, Sealants, Glazing and Waterproofing: Fourth Volume, ASTM STP 1243, David H. Nicastro, Ed., American Society for Testing and Materials, Philadelphia, 1995.

ABSTRACT: In today's commercial construction and renovation markets, the building owner desires a structure that remains both free of leaks and aesthetically pleasing for many years. It is assumed by many designers that within the group of high performance sealants "a silicone is a silicone", indicating that a sealant in a generic class carries with it both the positive and negative characteristics associated with that class. Silicone is generally considered the premium performing material, exhibiting excellent durability. However, occasional appearance problems, generically assigned to silicones, have swayed some specifiers away from silicones on some substrates. The contributors to appearance problems are many, and only a few are specifically related to the sealant used.

This paper describes several sealant related phenomena which can contribute to building aesthetic problems. These include sealant dirt pick-up, substrate staining and fluid migration into porous substrates. Each of these terms is defined in detail. Laboratory and weathering tests are described which indicate how quickly and how severely aesthetic problems develop. Results are reported for sealants representative of the following types: plasticized silicones, unplasticized silicones, unplasticized surface-modified silicones, plasticized polyurethanes, unplasticized polyurethanes and plasticized modified polyethers. The influences of local environment, joint design and application technique are also discussed.

KEYWORDS: aesthetics, staining, dirt pick-up, fluid migration, sealants

[1]Technical Service and Development Chemist, Scientist, Development Engineer and Global Technology Manager, respectively; Sealants Science and Technology, Dow Corning Corporation, Midland, MI 48686.

DIFFERENT TYPES OF BUILDING FACADE AESTHETIC PROBLEMS

In highly polluted areas or with substrates particularly prone to discoloration, the appearance of a building can degrade with time. The building may have vertical streaks on the substrate, the expansion joints may be blackened, or the entire facade may be covered with a film. All of these phenomena have been generically described as staining. To understand and address the causes, it is important to first separate the different types of aesthetic problems.

Substrate Staining

Discoloration of the substrate adjacent to a sealant may be attributed to airborne contaminants depositing on areas of the substrate where fluid has migrated from or has been washed down from the sealant. In the absence of the fluid, the contaminants would eventually erode and be washed away from the substrate. Substrate staining can appear as dark, vertical streaks on building facades or as darker areas surrounding both the horizontal and vertical joints.

Dirt Pick-up

Dirt pick-up describes the darkening of a joint due to airborne contaminants depositing on the sealant surface. The contaminants either form a thick, oily film on the sealant or are imbedded into the sealant. In either case, the particulate matter on the surface of the sealant is not generally removed by natural forces, such as sunlight and rain. In most cases, the joints are darkened or lightened (depending on the type of contamination and on the sealant color) obscuring the original color of the sealant. For example, a white paneled building may appear as a large checkerboard if the sealant is darkened by airborne contaminants. Dirt pick-up contributes to the above-described substrate staining by providing a reservoir of contaminants that can be washed down over the substrate forming dark streaks.

Fluid Migration

Migration of fluids (plasticizers, unreacted polymers, anti-oxidants and other liquid additives) out of a sealant and into a porous substrate causes a dark band around the expansion joint or weatherseal. The fluid soaked band may extend only a few millimeters or several centimeters from the joint depending on the amount and viscosity of free fluid in the sealant. Fluid migrating into the stone does not usually attract and hold contaminants, but makes the dry substrate appear wet in those areas and can cause differential wetting of the substrate. That is, the fluid can waterproof the substrate in the area adjacent to the sealant and make it appear dry when the rest of the substrate is wet.

Dirt / Residue Rundown

Even in the absence of sealant, vertical streaking can be caused by dirt collected on a roof or other flat area being channeled down the side of a building by water run-off. Residue rundown can also be caused by materials leeching from the substrate or other non-sealant building materials (i.e., efflorescence from masonry, oxides from metals, decomposition products from neoprene spacers, etc). Generally, the residue can be removed with cleaning without permanent marring of the substrate. Some residues, such as weld splatters and rust, can be permanent[1].

FACTORS RELATED TO BUILDING AESTHETIC PROBLEMS

Environment

In regions where air pollution is high, such as airports or downtown areas of large cities, all buildings get dirty. Air pollution is typically comprised of hydrocarbons generated by automobile engine or heating unit combustion or the reaction product of hydrocarbons and nitrogen dioxide (smog)[2]. These hydrocarbons act as carriers for solid particulate pollution. When these airborne contaminants are highly concentrated in the air they can form a visible film over the entire building exterior regardless of the sealant used.

A dusty job site will cause dirt to become imbedded in the sealant before it is completely tackfree. No amount of washing will remove that dirt. In areas where there is high ultraviolet exposure, such as the Southern exposure of buildings, organic sealants may degrade enough that dirt is sloughed off along with the top exposed layer of the sealant. Unfortunately, this is also a prime condition for organic sealant reversion or hardening[3,4] to occur.

Joint Design

Uncontrolled water channeling can cause rundown patterns to form even in the absence of a sealant[5] (defined as residue rundown in the previous section). Improperly formed weep holes, flashings and copings along with sloped roofs can contribute to dirt rundown problems. Particularly conducive to vertical rundown streaks on buildings is the use of joints that are "V"-shaped.

Application Technique

Overapplied sealant, even when wiped off, can leave a residue on the substrate that degrades very slowly. This residue is susceptible to dirt pick-up, but can be removed by mechanical abrasion or by treatment with a depolymerizing poultice. Overapplied primer can waterproof areas making them look as if fluid has migrated into the substrate when dry[6] or cause a "dry-look" when the substrate is wet. Some primers also yellow upon weathering. However, the primer will eventually degrade with time. Tooling a sealant with solvents or soapy water can contribute to dirt pick-up[7] and potentially cause streaking on glass or other surface-finished substrates. A sealant in contact with an incompatible material can discolor.

Sealant Condition / Composition

An out of shelf-life sealant gives an undercured material with a soft and / or tacky surface where dirt can be imbedded. A change in sealant color can also be caused by surface chalking or by pigment instability. A poorly curing sealant has more loose polymer that is free to migrate into stone or be washed over the substrate. A two-part sealant that has not been mixed in the proper ratios of curing agent to base or has been mixed inadequately can also result in an undercured sealant. A highly plasticized sealant has more fluid that is free to migrate out of the sealant and into adjacent porous substrates. Free fluid can also cause dirt pick-up and staining problems, particularly if the molecular weight is high enough to hold dirt.

COMPARATIVE STUDY OF THE AESTHETIC PERFORMANCE OF SEALANTS

Weathering in an Industrial Environment

Weathering at a 45 degree angle to horizontal--The aesthetic performance of twelve weatherproofing sealants were compared by exposing them in an industrial environment. These sealants were selected from the generic types of plasticized silicones, unplasticized silicones, unplasticized surface-modified silicones, plasticized urethanes, unplasticized urethanes and plasticized modified polyethers. The majority of sealants in this study are one-parts with a two-part urethane and a two-part unplasticized surface-modified silicone also included. The sealants were installed into composite panels consisting of white Italian marble, fiber reinforced cement panel (unpainted) and reflective glass. The composite was arranged to have the marble in the upper right corner, the fiber reinforced cement panel in the upper left corner, and the reflective glass across the bottom. The joint dimensions were 0.025 m width by 0.006 m depth. The entire composite measured 0.25 by 0.25 m and was positioned at a 45 degree angle facing South in Midland, Michigan. Positioning the samples at a 45 degree angle to horizontal accelerates both the dirt deposition phenomena and the sealant degradation rate. Samples were prepared and cured four days in a laboratory, then taken outdoors to be exposed within the fences of a chemical plant and less than one kilometer from a second chemical plant.

The composite substrates (Figure 1) were selected on their ability to show fluid migration (marble and unpainted fiber reinforced cement panel) and staining (reflective glass). Only the extent of fluid migration into marble will be presented in the tables; the cement panels are used for visual illustration. After weathering ten months in Michigan, there was a range of dirt pick-up and staining performance within the subgroup of plasticized silicones (sealants 20-1, -3, -4, -5 & -6). Some of the plasticized silicones were showing moderate dirt pick-up on the sealant surface where others had heavy dirt pick-up. Likewise, some of the plasticized silicones streaked the reflective glass panel where others caused a heavy haze over the entire glass. The levels of fluid migration into the marble pieces also differed between the plasticized silicones. Fluid from some of the plasticized silicones penetrated only a few millimeters into the marble, where fluid from others penetrated more than 20 millimeters.

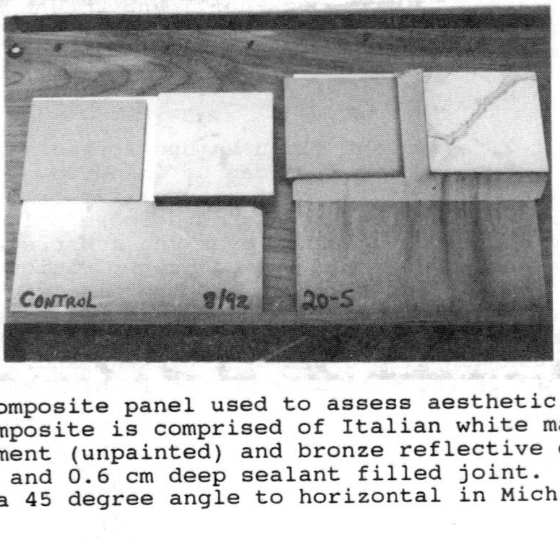

FIGURE 1 -- Composite panel used to assess aesthetic performance of sealants. Composite is comprised of Italian white marble, fiber reinforced cement (unpainted) and bronze reflective glass adjacent to a 2.5 cm wide and 0.6 cm deep sealant filled joint. The panels were weathered at a 45 degree angle to horizontal in Michigan.

After ten months, the unplasticized silicone (20-2) showed moderate dirt pick-up on the sealant and caused streaking and a slight haze over the reflective glass. The unplasticized surface-modified silicones (20-7 & -8) produced very little staining on reflective glass. The two-part, surface-modified silicone (20-8) had very little dirt pick-up on the sealant surface as well. The unplasticized silicones did not cause fluid migration into either the marble or fiber reinforced cement panels. Fluid from the modified polyethers (20-9 & -10) migrated into the marble pieces, the reflective glass panels were slightly streaked and the sealant surfaces remained clean. Neither polyurethane sealant (20-11 & -12) showed fluid migration into the marble. The two-part polyurethane (20-11) was slightly dirty and produced slight streaking on the reflective glass. The one-part polyurethane sealant (20-12) exhibited very little staining, virtually no dirt pick-up, but had pulled away from the marble. A high level of chalking was keeping the one-part urethane clean, but also caused severe surface crazing and would have appeared faded or lightened in any other color of sealant (Table 1).

TABLE 1-- Aesthetic Testing Results (45 degree angle, 10 months, Michigan)

Sealant	(1) Description	(2) Dirt Pick-up (1 - 5)	(2) Staining (1 - 5)	Fluid Migration into Marble (mm)
20-1	S/P/1	4	4	5-8
20-3	S/P/1	3	4	5-8
20-4	S/P/1	5	4	4
20-5	S/P/1	4	5	20-40
20-6	S/P/1	3-4	4	10-20
20-2	S/U/1	3	3-4	0
20-7	S/U/SM/1	3	2	0
20-8	S/U/SM/2	2	2	0
20-9	MPE/P/1	2	2	5-10
20-10	MPE/P/1	1	2	0-10
20-11	PU/P/2	2	2	0
20-12	PU/U/1	1	1-2	0
CONTROL	no sealant	...	1	0

(1) Sealant descriptions:
S/P/1 = one-part plasticized silicone
S/U/1 = one-part unplasticized silicone
S/U/SM/1 = one-part unplasticized surface-modified silicone
S/U/SM/2 = two-part unplasticized surface-modified silicone
MPE/P/1 = one-part plasticized modified polyether
PU/P/2 = two-part plasticized polyurethane
PU/U/1 = one-part unplasticized polyurethane sealant

(2) Dirt Pick-up and Staining ratings[8]:
1 - Sealant is as clean as before exposure and panel is as clean as control without sealant.
2 - Sealant is slightly dirty and panel is slightly streaked.
3 - Moderate dirt pick-up on sealant and panel is streaked.
4 - Dirt pick-up on sealant is visible from > 1.5 meters away and panel is covered with a hazy film.
5 - Heavy dirt pick-up on sealant and heavy haze over panel.

Unlike the silicones, the modified polyether and polyurethane sealants are comprised of degradable materials. This inherent tendency to decompose allows the sealant dirt pick-up and staining to be cyclical rather than cumulative. The poor, but intermittent, aesthetic performance of the organic based sealants was evident after weathering three months in Michigan (Figure 2a, Table 2). The modified polyethers (20-9 & -10) caused either streaking or hazing over the reflective glass, but the sealant surfaces stayed relatively clean. After weathering 10 months in Michigan, both the modified polyether sealant surfaces and reflective glass were relatively clean (Figure 2b, Table 1). A combination of heavy rains and increased sunlight from April to June contributed to the degradation of the fluid that had previously streaked the glass panel. However, fluid that migrated into the marble remained after 10 months of weathering making it a more long-term rather than cyclical effect. The one-part urethane sealant (20-12) showed only slight streaking and very little dirt pick-up after weathering three months and after ten months as reported above. Main contributors to the good aesthetic properties of the one-part urethane is its free chalking surface and lack of plasticizer. The two-part polyurethane sealant (20-11) caused both streaking and hazing on the glass panel and the sealant surface was significantly soiled after three months, but also returned to a relatively clean state after ten months.

(2a)

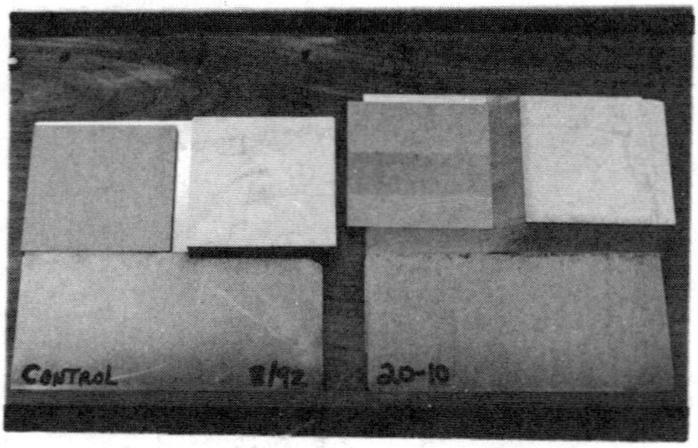

(2b)

FIGURES 2a & 2b -- Comparison of staining and dirt pick-up of plasticized modified polyether sealant (20-10) after weathering three months (2a) and ten months (2b), respectively. Note the presence of fluid streaks over the reflective glass after three months, but not after weathering ten months in Michigan. Also note that fluid that has migrated into the marble is still present after ten months.

TABLE 2-- Aesthetic Testing Preliminary Results (45 degree angle, 3 months, Michigan)

Sealant	(1) Description	(2) Dirt Pick-up (1 - 5)	(2) Staining (1 - 5)	Fluid Migration into Marble (mm)
20-1	S/P/1	3	3	2-3
20-3	S/P/1	3	3	2
20-4	S/P/1	3	3	2-3
20-5	S/P/1	2-3	3	9-12
20-6	S/P/1	2-3	3	8-10
20-2	S/U/1	2	3	0
20-7	S/U/SM/1	2-3	2	0
20-8	S/U/SM/2	1-2	2	0
20-9	MPE/P/1	1-2	3	2-4
20-10	MPE/P/1	1	4	4-5
20-11	PU/P/2	3-4	3-4	0
20-12	PU/U/1	1	2	0
CONTROL	no sealant	...	1	0

Weathering at a 90 degree angle to horizontal--Since building joints are rarely oriented at a 45 degree angle to horizontal, sealant samples were also exposed at 90 degrees to monitor realistic levels of dirt pick-up. The upper half of these 0.0025 m thick samples was masked and the lower half was exposed to the environment. After weathering for eleven months in an industrial location, the color change or difference in lightness between the masked area and exposed area was determined using a MacBeth Spectrophotometer according to ASTM Test Method for Dirt Collection Index (D 3719-87) (Table 3). The dirt collection index is a percentage of the lightness of the exposed sealant area to the lightness of the masked sealant area, where no dirt accumulation or change in sealant color corresponds to an index of 100.

The polyurethane sealants (20-11 & 20-12) had the cleanest surfaces with the unplasticized one-part polyurethane actually being lighter in color after exposure. The increased sealant lightness with exposure was due to the chalking of the one-part urethane surface. The two-part unplasticized surface-modified silicone (20-8) was also among the cleanest sealant surfaces. The group with surfaces slightly dirtier than the first group included three plasticized silicones (20-3, -5 & -6), one of the modified polyethers (20-10) and the one-part surface-modified silicone (20-7). The third group showed the most dirt pick-up and was comprised of two plasticized silicones (20-1 & -4), the unplasticized silicone (20-2) and one of the modified polyethers (20-9). Without the increased degradation rate afforded by exposure at a 45 degree angle, the dirt pick-up on the silicones was comparable to that of the organics. Exceptions to that generalization were the one-part polyurethane (20-12) where its free-chalking character allowed the surface to be replenished very rapidly and one of the plasticized silicones (20-4) that was markedly dirtier than all the other sealants. Also worth noting was that the darker sealants rated higher in the test. Perhaps the dark colors obscured the dirt that was collected on the sealant surface.

TABLE 3-- Dirt Collection Order (90 degree angle, 11 months, Michigan)

Sealant	(1) Description	Dirt Collection Index	Sealant Color
20-12	PU/U/1	100.21	white
20-11	PU/P/2	96.00	limestone
20-8	S/U/SM/2	94.75	off-white
20-6	S/P/1	93.91	gray
20-10	MPE/P/1	93.29	tan
20-7	S/U/SM/1	92.63	off-white
20-3	S/P/1	92.50	off-white
20-5	S/P/1	91.36	off-white
20-9	MPE/P/1	89.62	white
20-2	S/U/1	89.25	off-white
20-1	S/P/1	89.15	white
20-4	S/P/1	87.70	off-white

Accelerated Fluid Migration Testing

A semi-quantitative, accelerated method was used to measure the tendency of a sealant to discolor or darken porous substrates. The method can be applied to any porous substrate and in this case the substrate was one lot of white Italian marble. Three 0.013 m x 0.013 m x 0.025 m sample joints were prepared per sealant and allowed to cure for one month at 25 +/- 5C and 50 +/- 5% relative humidity (RH). The teflon spacers used to maintain the joint configuration during cure were then removed, but the closed cell backer rod was not removed. Each test piece was compressed 50% and placed into a 70C convection oven for one month. After removal from the oven, each test piece was broken in half and the cross-section distance that fluid migrated from the sealant into the marble was measured. There was considerable migration distance variance within different areas of the same test pieces due to porosity gradients within the stone, but results from one test piece to another were consistent. Accelerated migration results were compared to migration observed for the same type of marble substrate exposed to outdoor weathering (Table 4).

As expected, the plasticized sealants showed more fluid migration into the marble than the unplasticized sealants. Within the plasticized silicones, sealants 20-5 & 20-6 showed exceptionally high levels of migration where the entire test piece was saturated with fluid. The unplasticized silicones (20-2, -7 & -8) only darkened the area where the sealant contacted the marble. The organics all showed fluid migration, except for the one-part polyurethane (20-12). The accelerated method correlated well with the outdoor testing except for two sealants, where the plasticized silicone (20-4) showed fluid migration under natural weathering conditions, but not under the accelerated method and the plasticized urethane (20-11) exhibited migration under natural weathering, but not under accelerated test conditions. Substrate porosity, fluid viscosity and / or fluid branching along with the fluid decomposition rate each can play a role in the extent fluid will penetrate a substrate.

Extractable Fluid and Polymer Content

The amount of fluid or polymer that can be removed from the sealant by extraction techniques is a measure of what can potentially migrate out of the sealant. The extractable material can migrate into a porous substrate or it can lead to dirt pick-up and staining problems by setting on the sealant surface or running down the face of the building. If the molecular weight of the extractable material is high enough or the polymer is highly branched, it may never migrate out of the sealant due to chain entanglement. Functional material that is extractable shortly after the sealant is applied can eventually tie into the crosslinked network by a "post-cure"[4]. The extractable content also can change over the sealant service life by loss of fluid at the air / sealant or substrate / sealant interface.

As a way to compare the relative amounts of extractable material in each sealant, 0.013 m x 0.013 m x 0.025 m samples of the sealants on glass were cured for 6 weeks at 25 +/- 5C and 50 +/- 5% RH, then small pieces of cured sealant were heated under reflux in toluene for 24 hours. The extractable content of each sealant is reported (Table 4). With the exception of one of the plasticized silicone sealants (20-4), the extractable content per silicone sealant correlated well with the amount of fluid migration observed in both the accelerated and outdoor exposure tests. Due to the high reflux temperature of the solvent selected, the organic based sealants' extractable content was higher than that of the more heat-resistant silicones. Therefore, the extractable contents of sealants 20-9 through 20-12 were high relative to the amount of fluid migration exhibited.

TABLE 4-- Natural Fluid Migration, Accelerated Fluid Migration and Extractable Content

Sealant	(1) Description	(3) Natural Migration (mm)	Accelerated Migration (mm)	(4) Extractable Content (+/- 1%)
20-1	S/P/1	5-8	3	12
20-3	S/P/1	5-8	8	14
20-4	S/P/1	4	0	35
20-5	S/P/1	20-40	20+	23
20-6	S/P/1	10-20	20+	19
20-2	S/U/1	0	0.5	6
20-7	S/U/SM/1	0	0.5	7
20-8	S/U/SM/2	0	0.5	4
20-9	MPE/P/1	5-10	5	26
20-10	MPE/P/1	0-10	10	32
20-11	PU/P/2	0	4	30
20-12	PU/U/1	0	0.5	30

(3) Natural migration into white Italian marble was observed after 10 months weathering at a 45 degree angle to horizontal (Michigan)

(4) Extractable contents of sealants 20-9 through 20-12 were increased by high temperature conditions.

COMPARATIVE MECHANICAL PROPERTIES

The cleanest sealant in the outdoor weathering aesthetics study was the one-part polyurethane (20-12). A very close second was the two-part unplasticized surface-modified silicone sealant (20-8).

However, the aesthetic performance of a sealant is only part of the performance picture. A sealant must still be able to keep a building free of leaks for many years. In order to accomplish this, the sealant must be able to move with the building and not suffer long-term ageing or reversion effects. After three months curing at 25 +/- 5C and 50 +/- 5% RH a comparison of the initial mechanical properties revealed the one-part unplasticized polyurethane (20-12) and one of the plasticized modified polyethers (20-9) to have the highest 100% modulus values (Table 5). That is, at 100% extension, the polyurethane and the modified polyether are resisting extension with at least 50% more force than the other sealants. That resistance is transferred to the bond line and can ultimately lead to adhesive failures, especially if the sealant hardens with weathering.

TABLE 5-- (5) Mechanical Properties of Sealants after 3 months cure at 25 +/- 5C and 50 +/- 5% RH

Sealant	(1) Description	100% Modulus (kPa)	50% Modulus (kPa)	Break Elongation (%)	Break Stress (kPa)	Durometer (type A)
20-1	S/P/1	752	503	410	1620	38
20-3	S/P/1	931	600	391	2365	35
20-4	S/P/1	303	179	558	952	21
20-5	S/P/1	324	228	692	1455	18
20-6	S/P/1	621	455	350	1248	28
20-2	S/U/1	290	228	891	862	21
20-7	S/U/SM/1	531	393	530	1806	27
20-8	S/U/SM/2	579	441	420	1455	30
20-9	MPE/P/1	1172	655	214	2041	36
20-10	MPE/P/1	476	290	338	952	21
20-11	PU/P/2	731	455	238	1317	26
20-12	PU/U/1	1083	786	342	1476	50

(5) Test conditions: Sealants cured 3 months at 25 +/- 5C and 50 +/- 5% Relative Humidity

Sintech tensile tester, no extensiometer, 3.175mm x 50mm test area, dogbone, 500 mm / min crosshead speed

SUMMARY

The overall selection of a sealant should be based on a combination of weatherproofing performance and aesthetic considerations. Factors including the sealant's movement capability, adhesion to a building substrate and long service life should be included in every sealant selection process, where attention to appearance becomes an important consideration only when specific conditions lead to anticipated problems.

The majority of buildings existing today do not have an aesthetic problem, regardless of the sealant used, either because of their location, good design or because of the construction materials used. Heavy air pollution, poor building design, applicator error and other non-sealant related factors can contribute to the degradation of a building's appearance. The type of sealant used becomes important to aesthetics when the substrate is sensitive to staining or fluid migration, when air pollution levels are higher than normal and when chalking can change the sealant color. This paper included laboratory tests that are good predictive tools to indicate if staining or fluid migration can be potential issues in sealant selection.

The weathering and laboratory tests results showed that there are differences in aesthetic performance within the generic classes of "silicone" and "organic". The results also revealed that problems of fluid migration, dirt pick-up and / or staining existed for some sealants in each class. However, there were also both silicone and organic based sealants that had exceptionally good aesthetic performance.

REFERENCES

[1] Klosowski, J.M. Sealants in Construction Marcel Dekker, Inc. New York and Basel, 1989, Ch. 4, pp. 188 - 190.

[2] The World Book Encyclopedia Vol 6 (E), 1988 ed. Environmental pollution. World Book, Inc.

[3] Bridgewater, T.J., and Carbary L.D. "Accelerated Weathering and Heat Stability of Various Perimeter Sealants", Science and Technology of Building Seals, Sealants, Glazing and Waterproofing: Second Volume, ASTM STP 1200, Jerome M. Klosowski, ed., American Society for Testing and Materials, Philadelphia, 1992. pp. 48 - 50.

[4] Wolf, A.T. "Studies of the Ageing Behaviour of Gun-grade Building Joint Sealants - The 'State-of-the-Art'", Polymer Degradation and Stability, Elsevier Science Publishers Ltd. England, **23** (1989) pp. 138 - 141.

[5] reference 1, p. 194.

[6] reference 1, pp. 195 - 197.

[7] Gorman, P. "Sealant Lessons from Field Experience", Caulks and Sealants Short Course II. The Adhesives and Sealants Council, Inc. Washington, DC. Sept 16 - 19, 1990. Section VII p. 4.

[8] Dirt pick-up and staining ratings were adapted from those of US patent 4,695,603 p. 8 lines 41 - 52.

James C. Myers[1]

PERFORMANCE OF BAND-AID SEALANT JOINTS

REFERENCE: Myers, J. C., **"Performance of Band-Aid Sealant Joints,"** Science and Technology of Building Seals, Sealants, Glazing and Waterproofing: Fourth Volume, ASTM STP 1243, David H. Nicastro, Ed., American Society for Testing and Materials, Philadelphia, 1995.

ABSTRACT: This paper examines the performance of band-aid sealant joints through tensile and cyclic laboratory tests and finite element analysis. The testing includes joints with different bond breakers and band-aid shapes. The paper compares the results of this research with that of previous studies on performance of butt and fillet sealant joints. The results of this study provide a rational basis for designing and constructing band-aid sealant joints. The paper recommends design guidelines for these joints.

KEYWORDS: sealant joint design, sealant joint stresses, joint shape, movement capability, bond breakers, fillet joint, band-aid joint, butt joint.

The common sealant joint shape used in exterior walls of buildings is a butt joint, optimally with a 2:1 width-to-depth ratio and concave faces. The joint works well when the joint surfaces are parallel and gapped apart sufficiently. In some cases, constructing butt joints is not feasible for practical or economic reasons, and alternate joint shapes are needed. One such common situation occurs when the substrates are perpendicular and not gapped apart. In this case, one alternative is to create a fillet (triangular) sealant joint. I previously studied this joint shape and developed design guidelines for fillet joints [1 and 2]. Another common condition sealant designers and installers face is when the substrates are too shallow for a butt joint or when the parallel surfaces are not gapped apart sufficiently for a butt joint. One solution for this problem is to create a band-aid sealant joint over the gap. Some people prefer to call these seals bridge joints, since they span over the joint. Band-aid joints have been used on numerous occasions, but little guidance is available for designers as to the optimum

[1] Senior Project Manager, Simpson Gumpertz & Heger Inc., 297 Broadway, Arlington, MA 02174

configuration and movement capability of band-aid joints. This paper answers the following questions that band-aid sealant joint designers face:

- What is the movement capability of a band-aid sealant joint?
- How does the magnitude and nature of the stresses in band-aid sealant joints compare to butt or fillet joints?
- What is the best shape for a band-aid sealant joint?
- What size and shape are needed for the bond breaker?

This study uses tensile tests, cyclic tests, and a finite element analysis of one common band-aid joint to address the above-noted questions. The paper also references past results [1 and 2] of similar tests and analyses for various butt and fillet sealant joints to make relative comparisons to band-aid joints.

LABORATORY TESTING

This section describes the materials and procedures used in the tensile and cyclic tests, and presents the results of these tests. All testing consists of applying displacements perpendicular to the length of the joint in the plane of the band-aid. This study does not consider the effect of longitudinal movement on the joints. The nominal strain in the band-aid joint is defined arbitrarily as the displacement of the joint divided by the width of the bond breaker.

Description of Test Samples and Procedures

All test samples are 50.8 mm (2 in.) long. Test samples include three general types of band-aid joints. The Type A and B band-aid seals contain a tape bond breaker, with different surface profiles. The Type C band-aids are similar to the Type A seals, except they contain a 6.4-mm (1/4-in.) thick foam bond breaker. Figure 1 depicts the details of each test sample. The width of the bond breaker or foam backer is 12.7 mm (1/2 in.) in all test joints, since this is the standard ASTM butt joint width for test samples.

I selected these shapes based on the following constructability considerations and general findings in the previous study as to optimizing movement capability:

- Edges of the joints are tapered, so that water readily drains off of them when they are applied horizontally.
- The previous finite element analyses [1] on Group I fillet joints (i.e., those with foam backing at the root of the joint) indicates that the exact shape of the edge of the band-aid sealant joint is not likely to be a significant variable. For simplicity, the testing evaluates only band-aids with straight sloped edges, rather than including convex or concave edges.

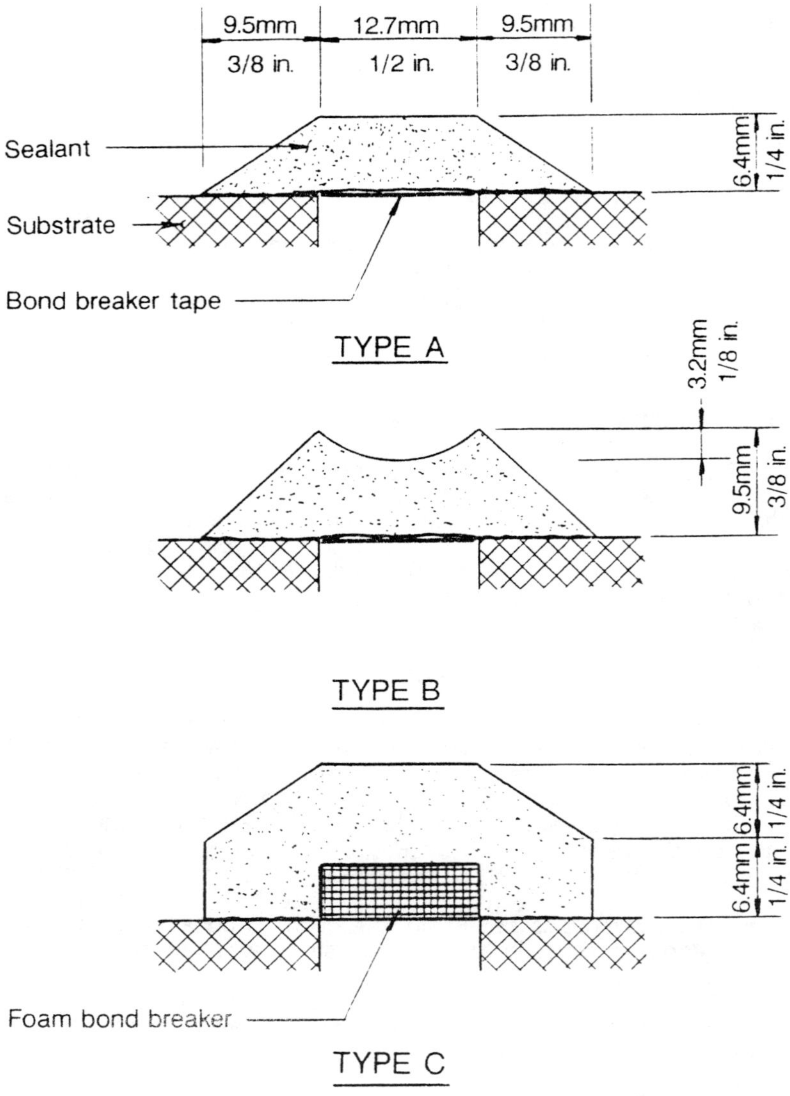

FIG. 1--Types of band-aid seals

- Minimum thickness of sealant is 6.4 mm (1/4 in.) to provide sufficient tolerances for expected variations in field construction. Thinner seals are likely to provide more movement capability, but cannot be constructed reliably. To control the magnitude of the testing program, the study did not include thicker sections, since they are less likely to have greater movement capability and may be more objectionable aesthetically.

- The study includes one joint, Type B, which contains a concave face, since butt joints with concave faces have greater movement capability than those with flat faces. A band-aid joint with a concave shape on the concealed face is not included, since foam backing materials needed to create such a shape are not readily available.

The primary focus of this study is to compare performance of various joint geometries, as opposed to various types of sealants. The test program includes only one sealant, a medium-modulus silicone with an ASTM C 719 rating of $\pm$ 25% on aluminum. This is the same brand, but different lot, of sealant used in the previous testing.

The displacement rate for all tensile tests is 12.7 mm/min (0.5 in./min). A custom jig holds the samples in the machine grips. The program includes three samples of each joint type.

The cyclic test procedure is similar to ASTM Standard Test Method for Adhesion and Cohesion of Elastomeric Joint Sealants under Cyclic Movement (C 719), except as modified below. The cyclic tests use several strains for each joint geometry to bracket the likely failure strain as follows:

- Joint Type A - 25%, 37.5%, 50%, and 75%

- Joint Type B - 50% and 75%

- Joint Type C - 50% and 75%.

The cyclic tests use a constant frequency (one-half hour per step), rather than a constant displacement rate. The program uses three room temperature cycles and four cycles with temperature changes. The samples cured for four weeks at standard conditions noted in ASTM C 719, but did not experience any water soak. After the week of compression in the oven, per ASTM C 719, they went through three room temperature cycles and then two cycles with temperature change. Due to lab scheduling, the samples were then compressed and stored at standard conditions for three weeks until the final two cycles with temperature change. The performance of the samples during the cycles, as noted below, did not justify any additional cycling for the purposes of this study. Mill finish aluminum angles form the substrates. The cyclic testing involves extending and compressing the joints using bolts with wing nuts and a series of shims manually inserted between opposing faces of the aluminum angles, since standard compression-extension machines for sealant joints cannot accommodate the band-aid geometry.

Tensile Test Results

Ultimate values refer to values at the point of maximum load. The average nominal ultimate strains ranged from 285% for Type A and 372% for Type B to 543% for Type C.

Previous testing of fillet joints [1] with the same silicone sealant found ultimate strains from 286% - 734% for Group I fillets (foam bond breaker) and from 234% to 472% for Group II fillets (bond breaker tape). Load elongation curves for band-aids and fillet sealants are similar.

The strain does not distribute uniformly along the band-aid joint based on visual observations of joint deformation in tensile tests, but concentrates at the concealed edges of the sealant along the bond breaker. All of the joints fail cohesively with splits along the edge of the bond breaker.

Cyclic Test Results

The passing strain criterion is similar to that specified in ASTM C 719. In previous testing [1], the results of the manual cyclic tests with square joints agree reasonably well with the manufacturer's ASTM C 719 automated test results. All three types of band-aid joints have passing strains of more than ±75%, i.e., none of the joints fail at the maximum test strain of ±75%. In fact, none of the joints show any appreciable distress or incipient failure at this strain. For this reason, additional cycles were not included in the test program. The Type C joint requires less force to cycle manually than does the Type A or B joint.

For reference, a square butt joint, which is the standard ASTM C 719 sample specimen, had a passing strain of ±25% in previous testing [1]. Other testing [2] of a 6.4 mm deep x 12.7 mm wide (0.25 in. x 0.5 in.) rectangular butt joint found that the joints did not pass ±50%, which was the smallest strain used to test the rectangular samples. Most of the Group I fillets do not fail at test strains of ±80% to ±100%, and most of the Group II fillets have maximum passing strains of between ±25% and ±50%.

NUMERICAL ANALYSIS

Analyses of stresses in sealant joints are complicated by the material nonlinearity of the rubbery sealant materials, the geometric nonlinearity due to large deformations, the incompressibility of the materials, the nonuniformity of the sealant cross-section, and the stress concentrating effect of the bond line. Reference 3 provides background information for readers unfamiliar with the finite element approach.

Assumptions and Procedures of Analysis

A finite element analysis based on linear elastic, small strain models examined the stress and strain distribution in the Type A band-aid sealant joint. Previous research [1] with both these models and nonlinear, finite strain (piece-wise linear) models did not find any significant differences with respect to comparative conclusions. Neither of these analyses are suitable for predicting absolute values of stresses. While many of the factors discussed below affect the absolute value of the results significantly, the effect on the comparative analyses, which are the purpose of this work, is less significant. The previous analyses [1] evaluated butt and fillet joints based on the same assumptions, except the earlier work used the MSC/pal2 program, while the analysis in this work used NASTRAN.

The analysis is two-dimensional, with the substrate displaced horizontally in the plane of the band-aid, perpendicular to the length of the joint. Plane strain conditions apply to the joint. Poisson's ratio is 0.49, and Young's modulus is arbitrarily 1 MPa (145 psi). The analysis does not consider the effect of time on the sealant behavior. The finite element analysis takes into consideration the sealant behavior due to presence of the bond line and joint geometry (stress concentrations) that are not present in dogbone tests.

Corners of the joints are points of stress singularity. The stresses that the program predicts at these locations are dependent on the size of the mesh in this region. To provide a uniform basis of comparison between the peak stresses in different shape joints, the mesh size near the corners duplicates the mesh size used in the previous analyses.

The modeling of any support or backing under the unbonded portion of the sealant changes the peak stresses. The analysis considers two cases, one with solid backing that prevents the band-aid from deflecting downward and one with no backing or restraint against downward deflection.

Results of Numerical Analyses

Based on the previous linear elastic, small strain analyses of joint extension, the peak stresses in a rectangular butt joint with a 1:2 width-to-depth ratio, i.e., a deep, narrow joint, are about ten times greater than in a Group I fillet joint with a concave concealed face. These are the maximum and minimum peak stress conditions found in the twenty-two joint configurations previously analyzed. For comparative purposes, these extremes are assigned dimensionless peak stress ranking values of ten and one respectively. The Type A band-aid joint analyzed in this research program has a peak stress ranking of 4.0 with solid backing and 2.7 without backing at the bond breaker.

For reference, Group I fillet joints have stress rankings up to 2.1. Group II fillet joints have stress rankings from 3.4 to 6.0. Rankings for square joints are 6.3. Rectangular joints with a 2:1 width-to-depth ratio have a 4.2 ranking with flat faces and a 2.6 ranking with concave faces.

The peak maximum principal stresses during extension of the Type A band-aid lie at the concealed face of the sealant just beyond the edge of the bond line. The general deformation and location of greatest deformation predicted by the analysis agree with behavior during testing. The predicted locations of peak stress correlate with the locations of failure initiation.

BAND-AID SEALANT JOINT DESIGN

Despite fairly widespread use of band-aid joints, little guidance has been available to designers faced with specifying them. This research into the behavior of band-aid sealant joints shows that these joints, when properly designed and installed, can perform as well as, or better than, butt joints and some fillet joints. This research provides guidance to designers on proportioning and movement capability of band-aid sealant joints and shows that many of the principles applicable to butt joint design also apply to band-aid joint design. As with butt and fillet joints, band-aid joints need adequate bond breakers to perform successfully.

Band-aid sealant joints perform better than square or rectangular butt joints and fall in the middle of the fillet joint spectrum. Band-aid joints have higher movement capabilities than butt joints and Group II fillet joints. Like most fillet joints, the band-aid joints do not experience direct compression. This avoids the substantial compression set inherent with butt joints and promotes increased movement capability.

A concave face on the band-aid tends to increase elongation capability slightly. Using a foam bond breaker, in place of tape, decreases the slope of the load/elongation curve and reduces the force that has to be applied in cyclic tests.

The designer should be sure that the substrate is stiff enough to withstand the "twisting" imparted by the band-aids. Very light gage metal copings may "dish" at the joints and pond water in an area that should be kept as dry as possible. Significant transverse slope in the coping can help overcome the twisting effect.

Joints with concave faces tend to "bulge" less when compressed, than do flat-faced joints. The aesthetics during compression can be a significant consideration in some situations.

Application Considerations

The band-aid geometry provides installation advantages and disadvantages over that of the butt joint. The substrate is more accessible to the applicator for cleaning and preparation.

Tooling of the joint is not difficult, but requires care to maintain straightness and consistent thickness. A plastic or metal putty knife can be cut to the desired band-aid profile easily. The most common source of problems occurs when the tool wanders from the joint centerline. With tapered edges, slight misalignment of the tool can alter the sealant thickness substantially over the edge of the bond breaker, which is the critical location for joint performance. Unlike with butt joints, there are not any substrate edges to guide the applicator's tool. Masking tape can help the applicator, but is not as positive as are edges of butt joints. Some people have used temporary foam tape spacers applied along the sides of the joints to facilitate sealant installation and tooling.

As with any sealant joint installation, pre-installation testing and mock-ups are essential. Mock-ups also allow evaluation of the applicators' technique.

DESIGN RECOMMENDATIONS

The following recommendations for design of band-aid sealant joints subjected to movement perpendicular to their length reflect the current knowledge regarding the behavior of sealant joints and may be revised as additional data become available.

1. Provide a bond breaker in all band-aid joints. Foam bond breakers increase the movement capability compared to tape bond breakers.

2. Divide the anticipated joint movement by the sealant's movement capability per ASTM C 719 to determine the appropriate width of the bond breaker, similar to the procedure used for sizing butt joints.

3. Maintain a minimum 6 mm (1/4 in.) sealant thickness over edges of bond breakers to allow adequate tolerance for field construction variability. Concave exposed faces improve performance, but are not essential.

ACKNOWLEDGEMENTS

Thanks to Simpson Gumpertz & Heger Inc. for use of their laboratory and computer facilities.

REFERENCES

[1] Myers, J. C. Behavior of Fillet Sealant Joints, Master's Thesis, Department of Materials Science and Engineering, Massachusetts Institute of Technology, Cambridge, MA, June 1989

[2] Myers, J. C. "Behavior of Fillet Sealant Joints," Building Sealants: Material, Properties and Performance, ASTM STP 1069, Thomas F. O'Connor, Ed., American Society for Testing and Materials, Philadelphia, 1990.

[3] Cook, Robert D., Concepts and Applications of Finite Element Analysis, Second Edition, John Wiley & Sons, New York, 1981.

Saad A. Abo-Qaidis[1] and Imad L. Al-Qadi[2]

RIGID PAVEMENT JOINT SEALANT EFFECTIVENESS IN REDUCING CHLORIDE INTRUSION

REFERENCE: Abo-Qaidis, S. A., and Al-Qadi, I. L., "**Rigid Pavement Joint Sealant Effectiveness in Reducing Chloride Intrusion,**" Science and Technology of Building Seals, Sealants, Glazing and Waterproofing: Fourth Volume, ASTM STP 1243, David H. Nicastro, Ed., American Society for Testing and Materials, Philadelphia, 1995.

ABSTRACT: To study the effectiveness of sealants in reducing chlorides and water intrusion into pavement joint openings, nine 122 x 91.5 x 20. cm concrete slabs were fabricated. One slab was cast without a joint and was used as a control; two were cast with non-sealed joints, and the other six were cast with joints which were sealed, after slabs were cured, with either type "A" or type "B" sealant, which are one-component rigid pavement joint sealants. Three dowel bars were installed below the joint at the mid-thickness of each slab, used for load transfer. Two types of dowels were used: epoxy-coated and uncoated. The joints were formed by installing removable styrofoam prior to concrete casting A crack was formed from the bottom of the joint to the slabs's bottom, through the cross section of the slab to simulate field conditions; construction paper was used for this purpose. To accelerate the possibility of dowel bar corrosion and to examine pavement joint sealant effectiveness in abating chloride intrusion, the slabs were exposed to alternate ponding cycles of 6% (by weight) sodium chloride solution (3-days ponding and 4-days dry). The corrosion progress was monitored by measuring corrosion current density (i_{corr}) using a Three-Electrode Linear Polarization (3LP) device. Also, chloride contamination was measured by collecting powdered concrete samples from each slab at four different depths to allow for chloride penetration profile assessment. The collected samples were analyzed to measure the percent of chloride diffused into the concrete and compared to background chloride percent in concrete. An analysis of results provided for the evaluation of i_{corr} and chloride contents over a period of 18 months. The results show that the unjointed slab has higher resistance to chloride penetration, followed by jointed slabs sealed with type "A" sealant, then jointed slabs sealed with type "B" sealant. In addition, the performance of epoxy-coated dowel bars in abating corrosion was better than that of uncoated dowels.

KEYWORDS: sealant, rigid pavement, concrete, chloride intrusion

The importance of rigid pavement joints arises from their ability to minimize stresses from environmental changes (temperature and

[1] Graduate Research Assistance, the Via Department of Civil Engineering, Virginia Tech, Blacksburg, VA.

[2] Associate Professor, the Via Department of Civil Engineering, Virginia Tech, Blacksburg, VA.

moisture) which result in expansion and contraction, to facilitate construction, and to control cracks by accommodating vertical and horizontal pavement movements. Concrete pavement joints may be classified as follows: transverse joints (expansion and contraction joints), longitudinal joints (center line and edge joints), and construction joints.

Expansion joints provide space for slab expansion and anticipated progressive horizontal movement from the gradual shoving of an adjacentslab. The joints are usually 19.1 mm to 50.8 mm wide and are spaced at intervals of 183 m. Contraction joints are designed to control random pavement cracking due to thermal shrinkage and warping stresses. These joints should be at least one fourth of the slab thickness deep, 6.4 to 19.1 mm wide, and spaced at 6.5 to 19.1 m. However, joints can be spaced as frequently as every 3.0 m if load transfer devices (dowel bars) are not used. Regularly spaced dowel bars are used generally at contraction and expansion joints to help transfer vehicular loads from one slab to the next.

Joint sealants may fail if one of the following problems occurs: adhesive failure, cohesive failure, extrusion or intrusion failure, and/or impregnation with incompressible materials. Failure in joint systems normally causes significant service life reduction of rigid pavements. A high percentage of deterioration in concrete pavements occurs at or near transverse contraction joints due to inadequately sealed or failed joints (Capas and Pennock, 1973).

The increasing use of deicing salts in the United States is also causing serious damage to pavement joint systems. Intrusion of salt and water through joint systems causes the dowels and tie bars to corrode (rust) because of chloride intrusion and the presence of oxygen and water. The expansive steel corrosion product (around six times the original volume) leads to increase internal stresses that ultimately lead to cracking and spalling (Mehta, 1986).

The rigid pavement materials around the joints are concrete and steel dowels. Of the two materials, steel possesses a stronger tendency for reactivity. The iron atoms inherently attempt to change to a more stable condition, preferably the original state from which the metal was extracted, predominately Fe_2O_3 (Hime and Erlin, 1987). To achieve a stable condition, the iron atom releases two electrons to form a positively charged iron ion. This ion, in concrete, combines with an oxide ion to form an iron oxide film layer. This layer gradually envelopes the steel creating a natural passive barrier between the iron and oxidizing reagents. The passive layer remains stable in the high pH concrete environment, until the surrounding is altered by external factors, mainly carbonation and chloride ions (Popovics and Sineonov, 1983).

Carbonation is the reaction of carbon dioxide with the free lime present in the concrete (Schiest, 1988). Chloride ions are more aggressive agents of deterioration of the passive layer. These ions permeate the sealant and the concrete to the steel reinforcement and directly affect the passive layer. The chloride ions react with the positively charged iron ions to form a product which has the capability to override the effect of the passive layer. Beyond the passive layer, the iron ions discharge the chloride ions and allow the chloride ions to cycle, so the corrosion process continues. At the cathode region, the free electrons from the iron react with water and oxygen to form hydroxyl ions, which cause destruction of the passive layer. At the anode region, the iron reacts with the hydroxyl forming iron oxide which migrates into the adjacent cracks and voids and reacts with additional oxygen to form expansive products. This creates an internal tensile pressure in the concrete and leads to cracks and delamination.

At present, pavement engineers can choose from hundreds of sealants. Extensive research has been conducted on the use of joint sealants in concrete pavements. However, the acceptable sealing of joints continues to be one of the most debatable issues in concrete pavement design and maintenance. The laboratory tests performed on sealants have concentrated on tensile strength, compression, bond, penetration, flow, stress relaxation, heat cycling, and environmental cycling (Collins, *et al.*, 1986, Jones, *et al.*, 1973, Thronton, 1978, Wolters, 1974, Spells and Klosowski, 1985, Tock, 1990, Evans and Romine, 1993).

Some studies investigated the change in physical properties due to environment. Tock (1990) reported a significant effect of temperature on silicone sealant properties. He also reported that the physical properties degraded significantly if sealant was immersed in acid water at a pH less than 2. High modulus sealant became embrittled, while low modulus material nearly dissolved to a viscous gum.

Other studies attempted to improve the testing technique for sealant evaluation. Spells and Klosowski (1985) improved a shear tester, the device used to simulate the vertical pavement deflection caused by trucks crossing the joint. Al-Qadi and Abo-Qudais (1994) developed a test method for evaluating sealant under simultaneous cyclic shear and normal deflections.

An examination of the available literature shows that sealant acceptance specifications require only laboratory testing that does not test for the effectiveness of sealants in preventing chloride and water intrusion into joint openings, which cause dowel corrosion that leads to concrete cracking, delamination and spalling. This may explain, in part, the poor field performance of some rigid pavement joints.

As part of a national pavement management system, a better understanding of pavement joint sealant performance is needed. It is well recognized that joint-sealant failures are caused by repeated vertical shear in the joint caused by heavy truck trafficking in addition to normal pavement joint opening and closing due to slab expansion and contraction caused by variation in temperature. However, there is no testing technique available to evaluate joint sealant in a manner simulating field loading and environmental conditions. As part of an ongoing research project to evaluate joint sealant effectiveness, a method to evaluate joint sealant under simultaneous dynamic shear and normal deflection was developed at Virginia Tech (Al-Qadi and Abo-Qudais, 1994). In addition, sealant effectiveness to abate chloride and water intrusion was investigated and is presented in this report.

RESEARCH PROGRAM

Test Sealants

Two types of rigid pavement joint sealants were evaluated in this study: Type "A" and type "B." Sealant "A" is a one-part, cold-applied material that readily extrudes over a wide temperature range. When it cures it is thought to produce a durable, flexible, low-modulus sealant. Because of its low-modulus characteristics and good extension/compression recovery (+100/-50 percent of original joint width), this sealant is easy to use with gunability at all temperatures. It also bonds to concrete without using a primer and is thought to have good weather resistance characteristics. Type "B" is a one-part sealant, cold-applied, that bonds to concrete by means of a primer which is applied to the joint concrete surface one hour before joint sealing.

Specimen Preparation

In January 1992, nine (122 cm X 91.5 cm, 20.3-cm-thick) concrete slabs were fabricated simulating jointed rigid pavement sections. After preparing wood forms, three dowel bars (45.7 cm long and 3.18 cm diameter) were installed at the mid thickness of each slab parallel to its length (Figure 1). One of the dowels was at the center of the slab, and the other two placed 15 cm from each edge. A type T thermocouple was connected to the center dowel bar in each slab to monitor slab temperature. Two types of dowel bars were used: epoxy-coated and uncoated (see Table 1). Each joint groove was formed by installing a pre-cut styrofoam piece, 6.4 mm thick by 50.8 mm deep, across the middle of the slab before casting the concrete. Also, a crack form the bottom of the joint to the bottom of the slab through the slab's cross section was formed, using construction paper, to simulate field conditions. When the forms were completely ready, the concrete was placed, and allowed to cure for 7 days before removing the styrofoam from the joint. The concrete physical properties are reported in Table 2. After seven days, the wood forms and the styrofoam pieces were removed. A backer rod, 9.5 mm diameter, was placed in the joint, 9.5 mm below the slab surface. The objective of the backer rod was to prevent three-sided adhesion between sealant and concrete, and also to form the desired sealant cross section. The sealant was then gunned to a depth of 3.2 mm below the slab surface to form a 6.4 x 6.4 mm cross section. The sealant was allowed to cure for one month before any sodium chloride solution ponding started. A 5-cm-high styrofoam dike was sealed at the edges of each slab to hold the sodium chloride solution during the ponding periods.

One slab was cast without a joint, using two uncoated and one epoxy-coated dowel bars, and was used as a control; two were cast with non-sealed joints, using two uncoated and one epoxy-coated dowel bars for each slab. The other six slabs were jointed; three of the joints were sealed with type "A" sealant and three with type "B" sealants. In each set of three, two slabs were cast with uncoated dowel bars and one with epoxy-coated dowel.

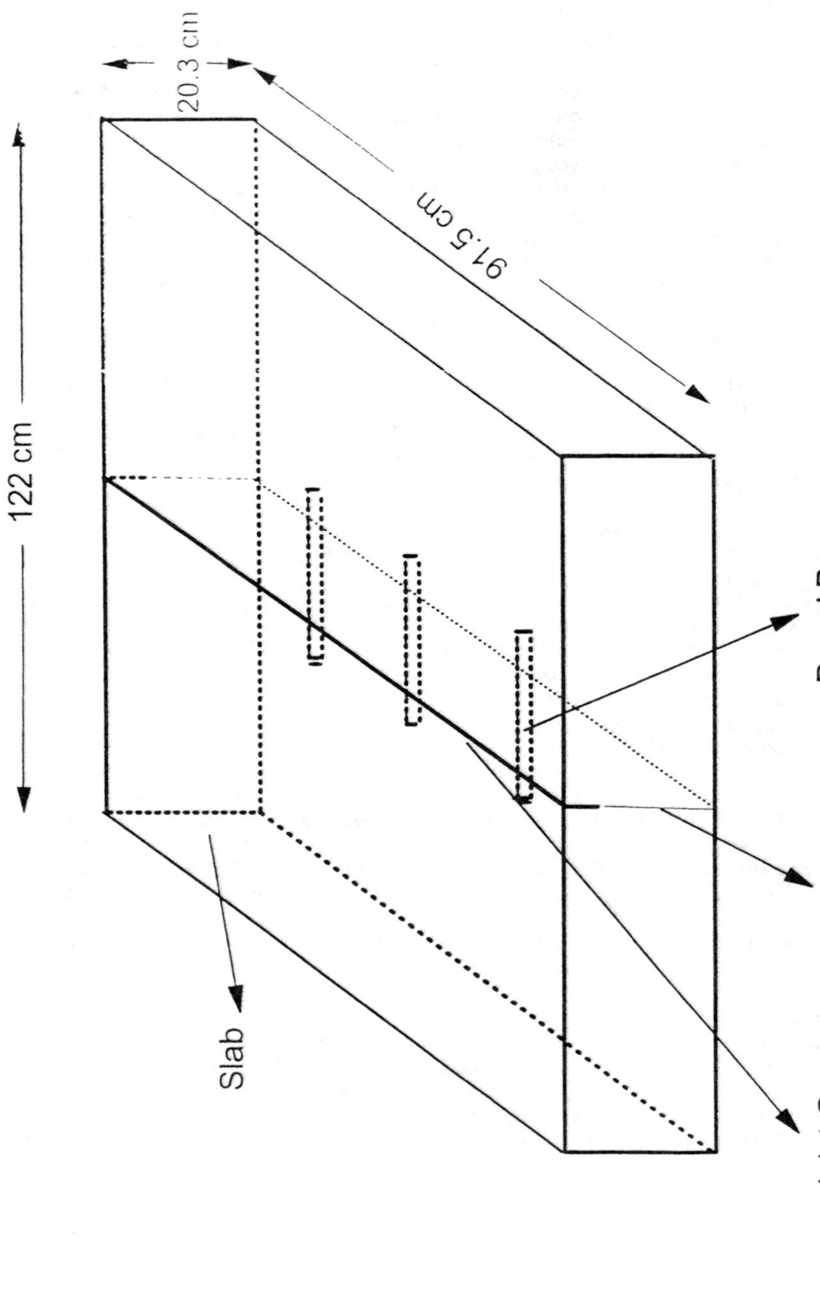

Fig. 1--Schematic diagram of fabricated slabs

TABLE 1--Joints, sealant and dowel bars used in different slabs.

SLAB NO.	JOINTED	SEALANT TYPE	TYPE OF DOWELS
1	YES	A	Uncoated
2	YES	A	Uncoated
3	YES	A	Epoxy Coated
4	YES	B	Uncoated
5	YES	B	Uncoated
6	YES	B	Epoxy Coated
7	YES	NONE	Uncoated and Epoxy Coated*
8	YES	NONE	Uncoated and Epoxy Coated*
9	NO	NONE	Uncoated and Epoxy Coated*

* The epoxy-coated dowel bar, in slabs 7 through 9, is located at one side of the slab, while the other two dowel bars are uncoated.

TABLE 2--Concrete mixture design and physical properties.

Coarse Aggregate	1025 Kg/m^3
Fine Aggregate	640 Kg/m^3
Type I Portland Cement	334 kg/m^3
W/C Ratio	0.49
28-Day Strength (Avg.)	26.12 MPa
28-Day Strength (STD)	0.51
Slump	6.9 cm
Unit Weight	2131 Kg/m^3
Maximum Aggregate Size	2.54 cm
Air Void Content	0.038

Laboratory Testing

To accelerate the corrosion of dowel bars and examine the sealant effectiveness in abating chloride intrusion, the concrete slabs were exposed to a 6% (by weight) sodium chloride solution (3-days ponding and 4-days dry). The corrosion progress was monitored by measuring corrosion current density (i_{corr}) using a Three-Electrode Linear Polarization (3LP) device, in which the rebar acts as a working electrode and a copper sulfide electrode pen cell (part of the 3LP device) is used as a reference cell. The polarization current is applied through a copper wire screening contained in a contact moist sponge. Corrosion current density (i_{corr}) was measured at both sides of each joint, at 2.5 cm from the joint edge. The corrosion current density was calculated using the Stern-Geary equation (KCC, 1990).

In addition, chloride contamination was measured by collecting powdered concrete samples from each slab at approximately 2.5 cm from the joint. At each location, four samples at 2.5 cm depth increments were collected, starting at 2.0 cm from the slab surface. This allowed for a chloride penetration profile assessment. The samples were collected using a rotary impact hammer with a 2.86 cm diameter masonry bit. The bit had a hole connected to a vacuum which was positioned high enough to suck the concrete powder formed near the bit head during drilling; this powder could be collected on a filter paper fixed at the top of the vacuum. The chloride content was measured using the specific ion electrode method developed at Virginia Tech (Herald *et al.*, 1992).In this method, an acid based solution was used to digest 3 grams of powdered concrete sample. Then, a dilute sodium chloride solution was added to the digested solution to reduce temperature effects and provide a background chloride ion concentration within the probe measurement range. The chloride ion concentration was then measured with a specific ion probe, which read in mV. The mV reading was converted to Kg/m^3 of chloride using a calibration equation for the probe used.

The slabs were exposed to 40 wet-dry cycles of sodium chloride solution. Corrosion current densities were measured after concrete placement and prior to sodium chloride solution bonding, after 8, 16, 24, 32, and 40 cycles and the last measurements were 8 weeks after the 40 cycles. Powdered concrete samples were also collected from each slab after 0, 18, 32, and 40 cycles of sodium chloride solution ponding, and the last samples were collected 8 weeks after the 40 cycles.

RESULTS AND DISCUSSION

Results of the measurements were presented as corrosion current density versus number of sodium chloride ponding cycles for each slab, and as chloride ion content versus depth for each slab at different numbers of cycles (see Tables 3 and 4, and Figures 1 through 6).

Corrosion Current Density Measurements

The corrosion current density (i_{corr}) measurements for uncoated dowel bars are presented in Table 3 and Figure 1. Type "A" sealant appeared to have higher resistance to salt infiltration than type "B" sealant. After 40 salt ponding cycles, the average corrosion current density for uncoated dowel bars in jointed slabs sealed with type "A" sealant was 43.43 mA/m^2 at a standard deviation of 4.12, while it was 48.34 mA/m^2 with a standard deviation of 3.67 for joints with sealant "B." Also, it can be noted that the unjointed slab experienced the lowest i_{corr} (3.25 mA/m^2), which indicates that sealants are not as effective as concrete in reducing chloride diffusion. On the other hand, unsealed jointed slabs experienced, as expected, the highest i_{corr} values over the evaluation period. The results demonstrated that sealants would reduce the intrusion of chlorides significantly, though less than what concrete would.

The same trend in corrosion rate density was observed for epoxy-coated dowel bars. Dowel bars in unsealed jointed slabs had the highest corrosion rate density followed by dowel bars in jointed slabs sealed with type "B" sealant, then by dowel bars in jointed slabs sealed with type "A" sealant. Again the lowest corrosion rate density was for dowel bars in unjointed slabs. Differences in corrosion rate densities between different slabs were not as pronounced as those for uncoated dowel bars because of the relatively small corrosion rate density of epoxy-coated dowel bars. It was noticed that in the case of epoxy-coated dowel bars, the zero voltage reading was fluctuating during measurements because of the effect of epoxy coating on the polarization process. Thus the i_{corr} measurements of epoxy-coated dowels were not as accurate as those for uncoated dowel bars.

Chloride Content Measurements

Chloride content measurements at the dowel bar level agreed with the i_{corr} results obtained (see Table 4). The chloride content was highest for unsealed slabs (average chloride content at 9.5 cm depth was 2.08 Kg/m^3 after 40 sodium chloride ponding cycles), followed by jointed slabs sealed with sealant "B" (1.72 Kg/m^3 at the same depth and number of cycles), then jointed slabs sealed with "A" sealant (1.17 Kg/m^3). Unjointed slabs had relatively small chloride ion content (0.91 Kg/m^3).

SUMMARY AND CONCLUSIONS

Nine concrete slabs were constructed to evaluate the effectiveness of joint sealants in reducing chloride intrusion. Different parameters were varied in the slabs including joint presence, dowel bar coatings, sealant type, and sealing of joints. Based on the information presented, the following conclusions are made:

Measurements of i_{corr}, for uncoated dowels, using a Three Electrode Polarization (3LP) device, indicated that the lowest corrosion rates were measured for dowels in unjointed slabs, followed by jointed slabs sealed with type "A" sealant, then by jointed slabs sealed with type "B" sealant. The rate of corrosion density was the highest for unsealed jointed slabs. The same trend has been noticed for slabs with epoxy-coated dowel bars. However the i_{corr}s measured in epoxy-coated dowel bars were not as accurate as those of uncoated ones.

Chloride content measurements, at different depths, supported the corrosion current density measurements. The same ranking of performance was obtained. None of the sealants failed during the study.

The study concluded that sealants reduce chloride intrusion significantly, but are less effective than Portland cement concrete. It is suggested that a careful selection of pavement joint sealant is fundemental. Also, epoxy coating inhibited the corrosion process.

TABLE 3--Corrosion current density for dowel bars in different slabs.

SLAB NO.	COATED DOWELS	0 Cycles		8 Cycles		16 Cycles		24 Cycles		32 cycles		40 Cycles		Last Measurement	
		i_{corr} (mA/m^2)	Temp. (°C)	i_{corr} (mA/m^2)	Temp. (°C)	i_{corr} (mA/m^2)	Temp. (°C)	i_{corr} (mA/m^2)	Temp. (°C)	i_{corr} (mA/m^2)	Temp. (°C)	i_{corr} (mA/m^2)	Temp. (°C)	i_{corr} (mA/m^2)	Temp. (°C)
1	NO	2.32	9.8	10.14	32.7	6.39	27.3	13.51	4.8	23.23	34.4	32.98	36.7	45.91	35.0
2	NO	4.75	11.2	20.43	32.1	22.74	25.7	17.53	5.0	31.39	38.9	53.88	38.3	35.00	37.8
3	YES	0.14	9.8	0.08	32.2	0.02	29.0	0.19	3.6	0.14	39.4	0.18	31.1	0.36	34.4
4	NO	15.38	11.1	18.19	30.8	13.60	24.9	18.18	3.7	26.42	40.0	39.82	33.3	44.10	33.9
5	NO	9.56	10.1	23.49	33.4	23.95	27.5	21.78	4.2	37.23	35.5	56.85	36.1	50.32	36.1
6	YES	0.06	9.9	0.12	32.1	0.11	28.7	0.23	4.5	0.15	36.7	0.33	30.6	0.48	30.0
7	NO	8.74	11.2	58.06	30.6	40.77	26.3	26.20	4.7	50.45	36.1	76.86	38.3	72.05	36.7
7	YES	0.02	11.2	0.13	30.6	0.33	26.3	1.87	4.7	1.32	36.1	0.83	38.3	0.85	36.7
8	NO	3.39	10.7	24.88	31.2	33.73	27.8	24.54	6.1	36.97	32.8	73.26	34.4	71.95	34.4
8	YES	0.07	10.7	0.35	31.2	1.28	27.8	0.92	6.1	0.70	32.8	0.57	34.4	0.58	34.4
9	NO	3.12	10.0	5.20	31.9	2.83	25.9	2.29	5.7	4.29	40.5	3.25	37.8	3.35	38.3
9	YES	0.04	10.0	0.18	31.9	0.06	25.9	0.04	5.7	0.04	40.5	0.06	37.8	0.04	38.3

* Coefficient of variation of i_{corr} ranges from 0.081 to 0.162.

TABLE 4--Chloride profiles of different slabs.

SLAB NO.	DEPTH (cm)	Cl^- (Kg/m³)				
		0 Cycles	18 Cycles	32 Cycles	40 Cycles	Last Measurement
1	2.0	0.54	6.79	12.69	15.78	14.99
	4.5	0.54	1.24	2.59	4.56	4.73
	7.0	0.54	0.72	0.74	1.54	1.55
	9.0	0.54	0.57	0.95	1.13	1.20
2	2.0	0.56	4.98	12.96	15.98	16.40
	4.5	0.56	0.92	1.38	1.79	2.04
	7.0	0.56	0.77	1.13	1.52	1.60
	9.0	0.56	0.64	1.05	0.99	0.94
3	2.0	0.45	9.91	13.41	14.80	14.12
	4.5	0.45	1.36	1.64	2.32	2.73
	7.0	0.45	0.66	0.79	1.42	1.38
	9.0	0.45	0.63	0.75	1.40	1.19
4	2.0	0.53	7.14	12.69	13.30	13.07
	4.5	0.53	1.04	2.36	1.80	2.04
	7.0	0.53	0.79	1.15	1.70	1.77
	9.0	0.53	0.69	1.00	1.68	1.58
5	2.0	0.42	9.69	10.72	14.99	14.30
	4.5	0.42	1.14	1.17	2.81	2.70
	7.0	0.42	0.70	0.86	1.71	1.51
	9.0	0.42	0.66	0.86	1.69	1.44
6	2.0	0.50	6.91	11.14	17.10	17.40
	4.5	0.50	0.72	1.65	3.22	3.06
	7.0	0.50	0.79	1.21	2.26	1.85
	9.0	0.50	0.67	1.07	1.80	1.66
7	2.0	0.62	8.88	7.80	13.53	14.61
	4.5	0.62	1.22	2.17	1.80	4.20
	7.0	0.62	0.95	1.33	1.70	1.96
	9.0	0.62	0.91	1.24	1.68	1.74
8	2.0	0.76	8.05	12.15	18.85	18.53
	4.5	0.76	1.39	1.47	2.99	3.23
	7.0	0.76	1.08	1.24	2.61	1.93
	9.0	0.76	0.99	1.21	2.48	1.49
9	2.0	0.57	9.86	16.18	19.91	19.66
	4.5	0.57	0.88	1.05	3.90	4.10
	7.0	0.57	0.71	0.83	1.23	1.11
	9.0	0.57	0.61	0.70	0.91	0.89

* Refer to Table 1 for joint type, sealant type, and dowel coatings.

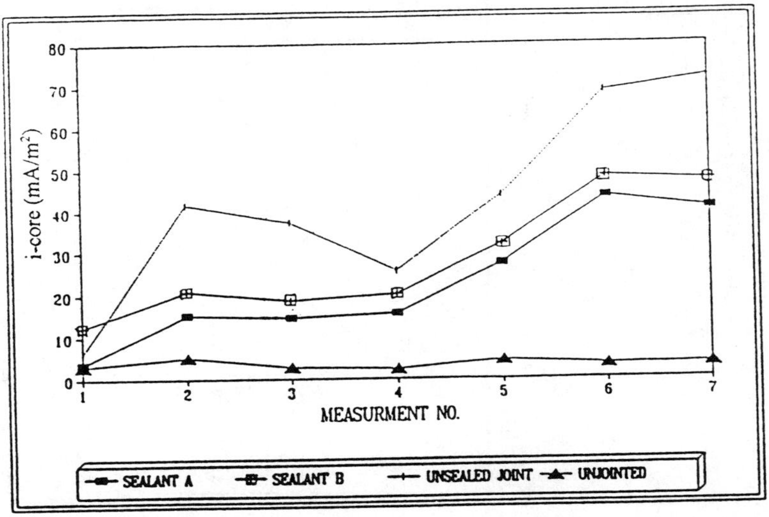

Fig. 2--Corrosion current density vs. number of cycles for uncoated dowel bars.

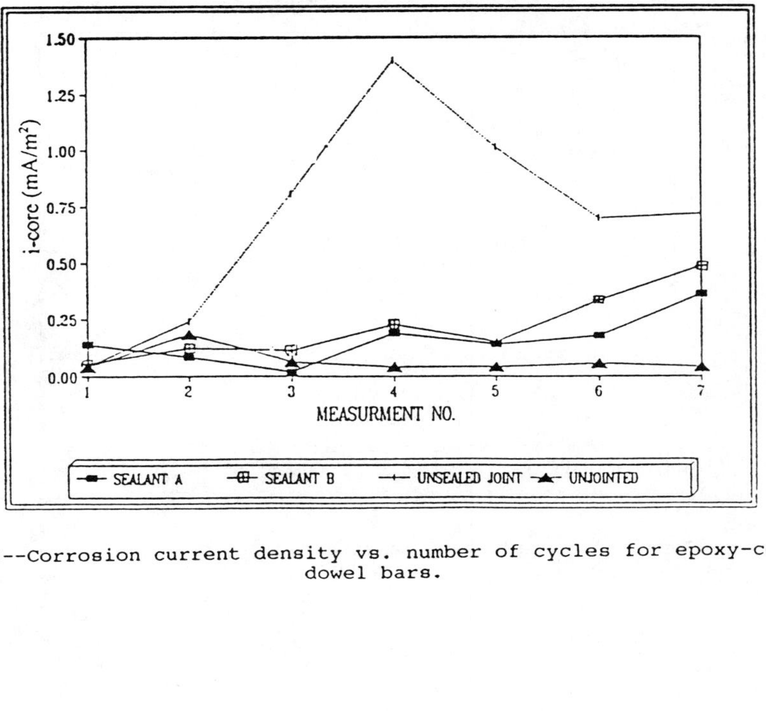

Fig. 3--Corrosion current density vs. number of cycles for epoxy-coated dowel bars.

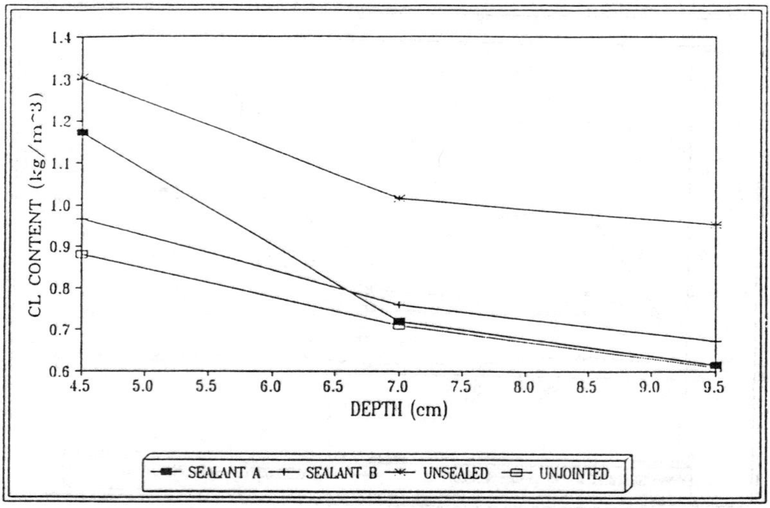

Fig. 4--Average chloride profile after 18 cycles of sodium chloride ponding.

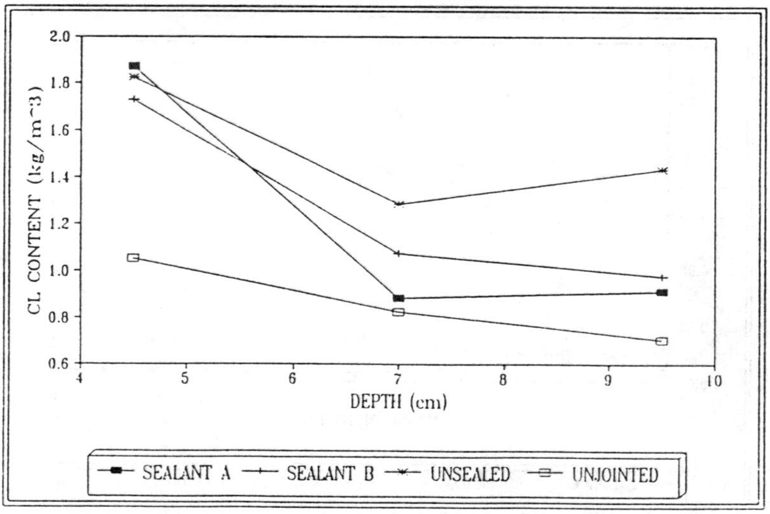

Fig. 5--Average chloride profile after 32 cycles of sodium chloride ponding.

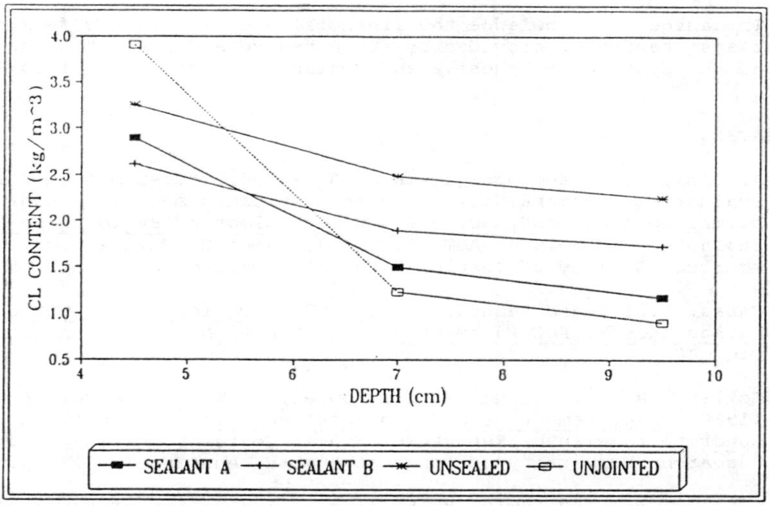

Fig. 6--Average chloride profile after 40 cycles of sodium chloride ponding.

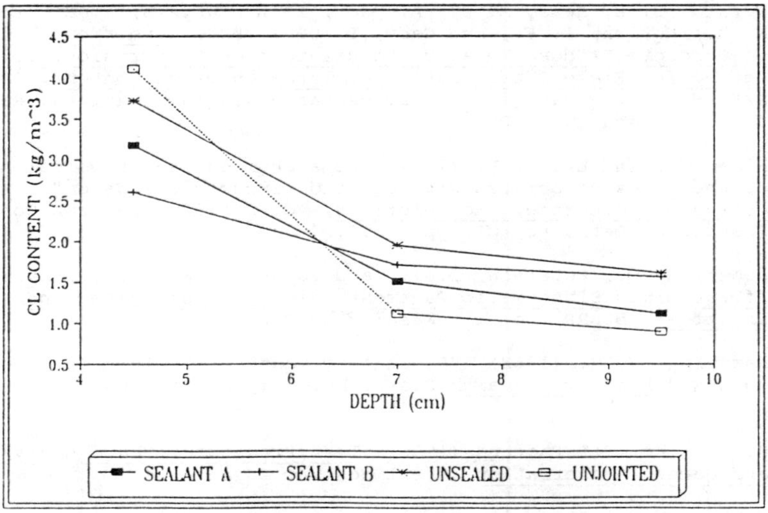

Fig. 7--Average chloride profile eight weeks after 40 cycles of sodium chloride ponding.

ACKNOWLEDGEMENT

The authors acknowledge the financial support, in the form of a student assistantship, provided by the Adhesive and Sealant Council, Inc. and the Center for Adhesive and Sealant Science at Virginia Tech.

REFERENCES

[1] Al-Qadi, I. L. and Abo-Qudais S. A. (1994), "Test Method for Evaluating Pavement Sealant Under Simultaneous Cyclic Shear and Normal Deflections", Science and Technology of Building Seals and Sealants, 3rd Volume, ASTM STP 1254, James C. Mayers, Ed., American Society of Testing Materials, Philadelphia, pp. 85-94.

[2] Capas, T. L., and Pennok, H. A. (1973), Design, Construction, and Maintenance of PCC Pavement Joints, NCHRP, No. 19, Washington, D.C, 40 p.

[3] Collins, A. M., Magnum, W. D., Fowler, D. W., and Meyer, A. H. (1986), Improvement Methods for Sealing Joints in Portland Cement Concrete Pavements, Report No. 385-1, Center For Transportation Research, University of Texas at Austin, Austin, TX, 132 p.

[4] Cook, J. P., and Lewis, M. R. (1967), Evaluation of Pavement Joint and Crack Sealing Material and Practices, NCHRP, No. 38, Washington, D.C, 37 p.

[5] Evans, L. D., and Romine, A. R. (1993), Innovative Material Development and Testing, Volume 4: Joint Seal Repair, SHRP-H-355, Strategic Highway Research Program, National Research Council, Washington D.C., 145 p.

[6] Herald, S. E, Henry M. B., Al-Qadi, I. L., Weyers, R. E., Feeney, M. A., Houlun, S. F., and Cady, P. D. (1992), Condition Evaluation of Concrete Bridges Relative to Reinforcement Corrosion, Volume 6: Method for Field Determination of Total Chloride Content, SHRP-S/FR-92-108, Strategic Highway Research Program, National Research Council, Washington D.C.

[7] Hime, W., and Erlin, B. (1987) "Some Chemical and Physical Aspects of Phenomena Associated with Chloride-Induced Corrosion" Corrosion, Concrete, and Chlorides, ACI SP-102, American Concrete Institute, Detroite, MI, pp. 1-12.

[8] Jones, G. M., Peterson, D. T., and Vayas, R. K. (1973), Evaluation of Preformed Elastomeric Pavement Joint Sealing System and Practices, NCHRP, No. 19, Washington D.C, pp. 32-41.

[9] Kuenning, W. H. (1962), "Laboratory Tests of Sealers for Sawed Joints," Highway Research Board, No. 211, Washington, D.C, pp. 1-12.

[10] K.C.C. Inc. 3LP Package (1990). Test Procedures, Data Analysis, and General Information, Sterling, VA, 43 p.

[11] Mehta, P. K. (1986), Concrete Structure Properties and Materials, Prentice Hall Inc., Englewood Cliffs, NJ, 548 p.

[12] Popovics, S., Simenov, Y., Bozhinov, N., and Barovsky, N. (1983), "Durability of Reinforced Concrete in Sea Water", Corrosion of Reinforcement in Concrete Construction, Ellis Horwood Ltd, New York, N.Y., pp. 19-38.

[13] Schiessl, P. (1988), Corrosion of Steel in Concrete, Champan and Hall, London, U. K, 102 p.

[14] Spells, S., and Klosowski, M. (1985) "Laboratory and Field Tests Can Predict Long Term Field Performance of Joint Sealant," Third International Conference on Concrete Pavement Design and Rehabilitation, Perdue University, West Laffeyette, Indiana, pp. 587-593.

[15] Thornton, J. B. (1978), "Investigation into the Performance of Rigid Pavement Joint and Material," Proceedings of International Conference on Concrete Pavement Design, Purdue University, West Laffeyette, Indiana, pp. 435-446.

[16] Tock, R. W. (1990), "Temperature and Moisture Effects on The Engineering Properties of Silicone Sealants", Building Sealant Materials Properties, and Performance, STP 1069 American Society of Testing and Materials, Philadelphia, PA, pp 167-173.

[17] Wolters, R. O. (1974), Field Evaluation of Joint Seal Material, Minnesota Department of Highway, Report No. ALA(2), Minneapolis, Mn, 19 p.

Timothy C. P. Lee[1]

POLYSULFIDE SYSTEMS FOR ENVIRONMENTAL PROTECTION

REFERENCE: Lee, T. C. P., "**Polysulfide Systems for Environmental Protection,**" Science and Technology of Building Seals, Sealants, Glazing and Waterproofing - Fourth Volume, ASTM STP 1243, David H. Nicastro, Ed., American Society for Testing and Materials, Philadelphia 1995.

ABSTRACT: Legislation in most advanced industrial countries is demanding more stringent protection of the water table. Chemical spillage and the diffusion of chemicals into the ground, leading to contamination of water, is being strictly controlled. A good example of these regulations is the containment of spillages of gasoline and diesel on gas station forecourts. Previous systems relied on the use of membranes (for example, heavy duty polyethylene sheeting). These are no longer thought to be reliable in the long term and more durable systems have been developed. These consist of fuel impervious concrete, sometimes in the form of hexagonal blocks, with joints sealed with fuel resistant sealants. The long term adhesion of the sealant/primer system to concrete immersed in fuels is one of the more difficult requirements of the specification. Other applications of polysulfide systems in pollution control are described.

KEYWORDS: polysulfide, sealants, coatings, pollution control, environmental protection

PROTECTING THE WATER TABLE

Throughout the industrial age it has been common practice to dispose of unwanted chemicals and waste oil and solvents in landfills or, in the case of liquids, by simply pouring them onto the land. For example, degreasing solvents used in the many thousands of engineering workshops around the world were often disposed of conveniently and cheaply by pouring them onto land on the factory site. It was assumed that the solvents evaporated, but in fact significant amounts permeated into the ground and thence into the aquifers.

A survey of the Coventry area in the UK showed levels of chlorinated hydrocarbons in ground water to exceed mandatory limits [1]. The contamination included fairly high levels of carbon tetrachloride, a solvent rarely used today, indicating slow permeation occurs over many years.

[1]Technical Director, Morton International Ltd, Polymer Systems, Sir William Lyons Road, Coventry, UK.

Such practices now are widely prohibited by laws which relate to both the deliberate disposal of liquids and to accidental losses. Thus it is a requirement that tank storage, above or below ground, has a secondary containment system, not only to contain small leaks from valves, but also to hold the complete contents in the event of catastrophic failure.

In countries such as the Netherlands, where the water table is only just below the surface, the problem is a critical one. There the control of spillage of liquids onto land has a high public awareness. It is the Netherlands and neighbouring Germany and the Benelux countries that have taken the lead in Europe in introducing legislation and specifications.

Across the USA there is equally intense activity aimed at controlling ground contamination, both by Federal Law and in individual states. High profile accidents, resulting in serious contaminations, have raised public awareness and accelerated the legislative processes.

Ironically it is the chemical and oil industries, the perceived villains, who are not only putting their houses in order, but are devising the systems which will help control and contain spillage. Some use engineering solutions, such as double wall tanks and tankers. Others are using the products of the chemical industry, such as polymeric membranes, coatings and sealants, to contain spillages.

POLYSULFIDE POLYMER TECHNOLOGIES

Polysulfide polymers were first synthesised in the 1920's and fully commercialised as solid and liquid rubbers in the 1940's. Polysulfide rubber was the first synthetic speciality rubber to be produced in the USA. It became successful because of its exceptional resistance to both aviation and automotive fuels. This resistance from the unique structural feature - the polymer contains 40% by weight of sulfur - has been exploited primarily in the design of sealants used in aircraft integral fuel tanks. All commercial and military aircraft use sealants and coatings based on polysulfide polymers.

Polysulfides resistance to weathering and their general durability led to the liquid polysulfide polymers being used in areas such as insulating glass and construction sealants, applications which generally do not exploit the polymers' inherent fuel and solvent resistance. However, with the need to develop technologies to control fuel, oil, solvent and chemical spillage, this property of liquid polysulfide polymers has finally found a major sealant use outside the aircraft industry.

Table 1 illustrates the comparative resistance of a number of synthetic rubber compounds to a range of fuels. The specific compounds used were standard ones for these particular rubbers and the swell resistance recorded was typical. The polysulfide compound used was based on a liquid polysulfide polymer LP-31, mixed with a medium thermal carbon black (30 parts per hundred by weight on the liquid polysulfide) and cured with manganese dioxide. With the exception of the fluorocarbon rubber, the polysulfide system exhibits significantly lower volume swell than the other rubbers.

The resistance to fluids has been exploited by the rubber industry for a number of years. Solid polysulfide rubber is used in the manufacture of printing blankets and rollers and in hose for the fuel and paint industries. The liquid polysulfide polymer systems cure into solid elastomers, which have the same structure and properties as their solid counterparts.

TABLE 1--Comparitive fuel resistance for rubber compounds

Rubber compound based on	% Vol change in fuel (3 days @ 23°C)			
	ASTM Fuel B	ASTM Fuel C	ASTM Fuel D	Alcohol Modified Fuel *
Chloroprene	66	109	90	72
Fluorocarbon	1.5	2.6	2.3	4.4
Fluorosilicone	15.2	17.8	15.9	19.5
Low nitrile	47	76.8	62	74
High nitrile	12	22.6	18.1	31.5
Epichlorhydrin	15	25	20.4	24.4
Polysulfide	3.1	11	7.9	9.3

* Iso-octane 48, Toluene 32, Ethanol 20

Even in high aromatic fuels, the volume swell is relatively low, as it is in the alcohol modified fuel. Table 2 gives volume swell data for the polysulfide compound in a range of modified fuels.

TABLE 2--Polysulfide coating volume swell

Fuel Composition (% by volume)			
Iso Octane	Toluene	Methanol	Volume Swell % 7 days at 23°C
60	40	...	7.9
45	45	10	17.2
42.5	42.5	15	17.6
40	40	20	16.6
50	30	20	9.3
75	15	10	3.3

The resistance of polysulfide polymers to fuels extends to a wide range of solvents and chemicals. The exceptions are chlorinated solvents (generally no longer used) and strong oxidising acids.

The structure of the polysulfide polymer also leads to other useful characteristics in properly formulated compositions:

- flexibility down to -45°C
- resistance to weathering (UV, ozone)
- resistance to long term water immersion
- low permeability to gases and water vapour.

The need to properly formulate has to be emphasised. In order to utilise the characteristics described above, the formulated system must contain a high level of the polysulfide polymer and should not have its performance diluted by over extension with plasticisers and fillers. For example, a typical aircraft fuel tank sealant has a polymer content of 60% or more.

APPLICATIONS FOR POLYSULFIDE POLYMER ENVIRONMENTAL TECHNOLOGIES

Coatings for Secondary Containment

Legislation in many countries requires that tanks, above or below ground, have secondary containment to control spillage. It is customary to put a bund wall (or dike) around the tank or tanks with a capacity to contain the total contents of the liquids stored. These are usually concrete or brick walls built on a concrete floor. Whilst adequate when first built, there is a tendency, with time, for brickwork to deteriorate or concrete to develop cracks. It has been proven that a cost effective way to make the containment effective once more is to overcoat the surfaces with a suitable membrane coating. A suitable coating has the following characteristics:

- flexible, to bridge over cracks
- durable, i.e. remain flexible and not itself crack, over a period of many years
- resistant for at least seventy-two hours to the fluids contained in the tank. (The seventy-two hours is required to take account of unnoticed spillages, over a weekend for example, plus enough time to pump spillage to another container).

A system which meets these requirements is a combination of liquid polysulfide polymer with a reinforcing carbon black cured with manganese dioxide. The inherent durability of the polysulfide polymer is enhanced by the presence of the carbon black, which supplies toughness and UV stability. After 10,000 hours under an 800W tungsten UV lamp such a coating exhibited no deterioration.

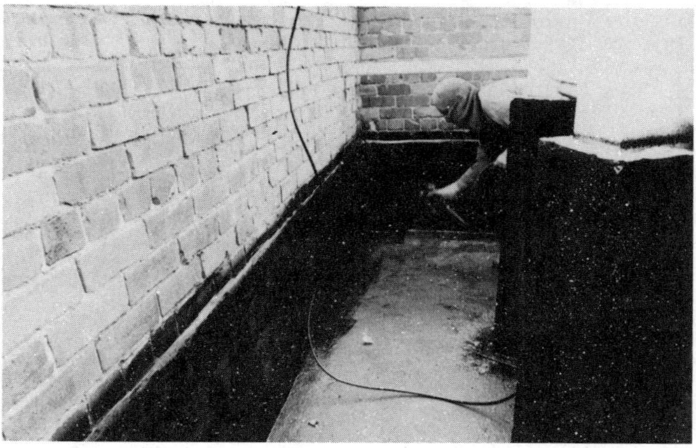

FIG. 1--Applying a polysulfide coating to a brick bund wall (dike)

Prior to application of the coating defects in the concrete or brickwork are repaired. Cracks are opened up and sealed with an elastomeric sealant with resistance to the stored fluid. It is normal to prime the surfaces with an epoxy type primer prior to applying the coating. The coating, which can be solvent free, is applied either using conventional airless or air assisted spray equipment or by brush (Fig. 1). The system takes about two hours to cure to a seam-free elastomeric membrane. Final coating thickness is typically 1mm to 2mm obtained with two coats. The properties of a polysulfide polymer coating are shown in Table 3. A volume swell of less than 150% after 72 hours is acceptable.

TABLE 3--Properties of a polysulfide coating

Property	Value
Hardness, Shore A	41
Tensile Strength, MPa	2.5
% Elongation at Break	490
200% Modulus, MPa	1.2
Permeability (gasoline) cm/sec	1.23×10^{-9}
% volume swell 72 hours at 23°C.	
ASTM Fuel D	15
Methanol	0.6
Toluene	120
MEK	62
ethyl acetate	42

A recent innovation in the USA is the development of a system to make environmentally secure earthen containment walls as a viable alternative to concrete or brick. This has particular use where existing tanks have no provision for containment.

The site is prepared by surrounding the tank(s) with an earthen dike to a height such that the volume of the resulting reservoir is sufficient to contain the contents of the tank(s). After ensuring the earthen walls and the floor are free of sharp intrusions, lengths of fabric which have been precoated with a carbon black reinforced polysulfide system, are laid down completely covering the earth. The overlapping seams are joined using a spray-applied polysulfide coating of the type described above. Finally, the whole area is overcoated with the polysulfide coating, giving rise to a seamless membrane (Fig. 2).

FIG. 2--Polysulfide coating over fabric on earthen dike.

Chemical Resistant Sealants

The use of slab concrete is widespread throughout the chemical and oil industries in transportation and in a large number of factory environments. It is normal for such slabs to have expansion joints at regular intervals. These are usually sealed to prevent rain water penetrating below the slabs. In today's pollution-sensitive world there is a new need - the slabs in critical areas have to be sealed with sealants resistant to fuel or chemicals to prevent penetration of liquids into the water table.

Chemical resistant sealants must have a high polymer content and a minimal amount of extractable ingredients such as plasticizers. They must adhere to the substrate (primers are used) and be sufficiently elastic to withstand movement. Collectively, these demands are a severe test for the formulators.

Sealants based on polysulfide polymers are finding applications in these industrial areas. For example, sealing hard-standing areas at airfields, where spillages of aircraft fuel and hydraulic fluids occur, is an area where polysulfide sealants have been specified for some time. The use of polysulfide sealants at fuel storage areas and refineries is growing in Europe.

Where the local regulations are particularly strict, the concrete may be laid over a heavy-duty polyethylene membrane. The concrete itself will be a grade with low permeability. In addition, large chemical plants can have many miles of concrete channels and ducting, for carrying services, cables and for run-off and containment of fire fighting water. The same stringent requirements for construction and sealing are applied to prevent permeation of contaminents into the water table.

Unlike for rubber, specifications and test methods for the chemical resistance of sealants are not widely available. In Germany, the Society of Industrial Construction has devised a specification which compares modulus and adhesion before and after immersion in the test fluid. The sealants committee for ISO (TC59/SC8) has the subject as a possible future work item. There appears to be a need to accelerate such activities in the light of tightening environmental legislation.

Sealants for Gas Stations

A special application for chemical resistant sealants has grown rapidly in Europe in recent years. A survey commissioned in Germany by the Association for Mineral Oil, Natural Gas and Coal (DGMK) examined ground contamination in the area of gasoline filling stations [2]. It is estimated that about 4000 tonnes of fuel are spilled every year at German gas stations, the majority of which evaporates. However, the survey concluded that the less volatile diesel fuel permeates into the ground to the extent "it has become a cause for concern". As a rough guide, one litre of diesel can spoil one million litres of water.

The problem is a serious one in areas where the water table is near the surface, as in the Netherlands. It was here in 1986 that it became mandatory to seal the immediate area of the pumps at filling stations and to control fuel spillages prior to safe disposal. A set of regulations were put in place, specifying impermeable concrete and a fuel and oil resistant sealing system. Initial trials compared polysulfide sealants with an absorbent sand/bentonite mixture, with the former prevailing [3].

A specification was soon devised for the whole technology by the Dutch organisation KIWA. The preferred flooring consists of precast concrete hexagonal blocks, with a density which makes them impermeable to fuel and oil. These are laid in a regular pattern with a uniform gap between blocks. Dry sand is blown into the gaps and compacted to an established depth. After priming the concrete the fuel and oil resistant sealant is applied (Fig. 3). To ensure the effectiveness of the flooring, which has to withstand almost continuous vehicular traffic, a high standard of workmanship is essential.

FIG. 3--Application of a polysulfide sealant at a gas station

The sealant/primer system is specified in the KIWA standard C50 [4]. It contains a number of standard sealant tests. However, it also includes tests related to the end use:

- Tensile adhesion and volume swell after immersion in standard petroleum and diesel fuels at 50°C.

- Resistance to shear. The test piece is subjected to 100 cycles through 100% extension in shear (simulating traffic movement).

Several proprietary polysulfide sealants manufactured in Europe meet the requirements of KIWA C50. It is estimated that about 1200 gasoline stations have been sealed with this technology in Holland. In Germany, which is committed to seal all its gasoline stations by 1996, 500 gasoline stations have been sealed so far.

The question is sometimes asked, what happens to the spilled fuels and oils now lying on the sealed surface? The answer lies in good housekeeping. The surfaces are washed down daily and the washings channelled into separators for later disposal.

CONCLUSION

The control of pollution in industrialised nations is a central part of the environmental strategy. The stringent standards represent a significant challenge to engineers, chemists and technologists. Fortunately, it is a challenge to which industry is rising, with the development of new technologies and the adaption of existing ones. With new chemistry and formulations, the long established polysulfide sealants and coatings fall into both these categories. The full benefits of their fuel, oil and chemical resistance are, for the first time, being fully exploited in civil engineering.

REFERENCES

[1] Lerner, D. N. and Tellam, J. H., Journal of the Institution of Water and Environmental Management, Vol. 6, February 1992, pp 28-37.

[2] Deutsche Wissenschaftliche Gesellschaft für Erdöl, Erdgas und Kohle, ev, DGMK, Steinstrasse 7, 2000 Hamburg, Germany. Project 422, January 1989.

[3] Deutsche Wissenschaftliche Gesellschaft für Erdöl, Erdgas und Khole, ev, DGMK, Steinstrasse 7, 2000 Hamburg, Germany. Project 422-2, February 1991.

[4] Keuringsinstituut Voor Waterleidingartikelen nv, KIWA, 2280 AN Rijswijk, Netherlands, Criteria No 50, Criteria for Channel Sealants, July 1989.

Anthony R. Fiorillo[1], Jackie R. Harding[2]

PRIMERLESS POLYSULFIDE BUILDING SEALANTS

REFERENCE: Fiorillo, A. R., and Harding, J. R., **"Primerless Polysulfide Building Sealants,"** Science and Technology of Building Seals, Sealants, Glazing, and Waterproofing: Fourth Volume, ASTM STP 1243, David H. Nicastro, Ed., American Society for Testing and Materials, Philadelphia, 1995.

ABSTRACT: Polysulfide polymers have a long history of success as the base polymer in building sealants, but these sealants usually require the use of primers to ensure good adhesion to the various substrates used in building construction. Usually different primers are required for porous and non-porous substrates.

Whenever primers are required to achieve good adhesion, there is always the risk that the primer may be improperly applied or that some areas may be skipped during the priming. There is also the need to ensure that the primer does not get on the building surface and cause discoloration.

Polysulfide sealants have now been developed that incorporate adhesion promoters that eliminate the use of primers. These sealants have been tested against ASTM C-920-87, Specification for Elastomeric Joint Sealants, as a class 25 building sealant on glass, anodized aluminum and concrete and have easily met all the requirements.

KEYWORDS: Polymers, Polysulfide, Sealant, Primer, Adhesion promoter

---

Most polymers used in building sealants do not have inherently good adhesion to all of the substrates commonly used in building construction. For this reason, it is usually necessary when developing a building sealant to develop a complete system that includes not only the sealant but any primers that may be necessary to give good adhesion to the particular substrates being used. This has been the practice with most

---

[1]Group Leader, LP Technical Service & Applications, Morton International, Woodstock, IL 60098

[2]Technologist, Polysulfide Applications & Technical Service, Morton International, Woodstock, IL 60098

sealant manufacturers for a number of years.

Polysulfide polymers have a long history of success as the base polymer in building sealants. These sealants also required the use of primers to assure good adhesion to the various substrates they were used with. Usually different primers were used for porous substrates and for non-porous substrates.

Typical primers used for polysulfide sealants include silanes, titanates and chlorinated rubbers for non-porous substrates and chlorinated rubbers or urethane polymers for porous substrates.

Whenever primers are required to achieve good adhesion, there is always the risk that the primer may be improperly applied or that some areas may be skipped during the priming, leading to premature sealant failure. Also, when applying primers, care must be taken to ensure that the primer does not get onto the building surface, since this will often cause discoloration on ageing.

Often times, in addition to using primers with sealants, the sealants will also contain adhesion promoters to help ensure good adhesion. Many types of adhesion promoters are available. Typically the proper adhesion promoter depends on the polymer being used and the substrates that the sealant must adhere to. The traditional adhesion promoters for polysulfide sealants are silanes, titanates, epoxy resins and phenolic resins.

Today, many new potential adhesion promoters are available to the formulator. By choosing the right combination of these materials, the probability of developing sealants with long term durability without the use of primers is greatly improved.

Polysulfide sealants have now been developed that do not depend on primers for their good adhesion. These sealants incorporate adhesion promoters that eliminate the need for primers for most applications.

The following formulation is for a two-part, manganese dioxide cured polysulfide sealant. This sealant incorporates a combination of adhesion promoters which give the sealant very good adhesion without the use of a primer.

Suggested Starting Formulation For a Two-Part Manganese Dioxide Cured Polysulfide Sealant

TABLE 1

Ingredient	Parts by Weight
Part A	
Polysulfide Polymer	100.0
Treated calcium carbonate	50.0
Pigment	10.0
Chlorinated plasticizer	60.0
Aluminum silicate	25.0
Phenolic adhesion promoter	5.0
Chain-stopper	0.1
Silane adhesion promoter	1.0
Part B	
MnO_2	10.0
Phthalate plasticizer	10.0
Accelerator	0.5

This formulation incorporates a phenolic resin as the adhesion promoter for porous substrates and a silane as the adhesion promoter for non-porous substrates. When tested against ASTM C 920-87 as a class 25 sealant for concrete, anodized aluminum and glass it easily met all the requirements of the specification without the use of a primer.

TABLE 2

<table>
<tr><td colspan="2">Sealant: Two-part MnO_2 cured</td><td colspan="2">Reference: 358-48</td></tr>
<tr><td>Type: M</td><td>Grade: NS</td><td colspan="2">Cure: 2 weeks at 25°C</td></tr>
<tr><td>Class: 25</td><td>Use: NT, M, G, A</td><td colspan="2">Primer: None</td></tr>
<tr><td colspan="2"></td><td>SPEC LIMITS</td><td>TEST RESULTS</td></tr>
<tr><td colspan="2">Rheological properties:
Horizontal at: 50°C
5°C
Vertical at: 50°C
5°C</td><td>
No deformation
No deformation
4.75mm max. sag
4.75mm max. sag</td><td>
No deformation
No deformation
No sag
No sag</td></tr>
<tr><td colspan="2">Extrusion Rate</td><td>45 seconds max.</td><td>29 seconds</td></tr>
<tr><td colspan="2">Hardness, Shore A: Original</td><td>15 to 50</td><td>25</td></tr>
<tr><td colspan="2">Effect of Heat Aging</td><td>10% max. wt.loss
Max. hardness 50
No cracking
No chalking</td><td>2.6% wt. loss
Hardness 28
No cracking
No chalking</td></tr>
</table>

TABLE 3

Sealant: Two-part MnO_2 cured		**Reference:** 358-48
Type: M	**Grade:** NS	**Cure:** 2 weeks at 25°C
Class: 25	**Use:** NT, M, G, A	**Primer:** None
	SPEC LIMITS	**TEST RESULTS**
Tack Free Time, Hrs.	72 max.	<24
Stain and Color Change (on mortar)	None, air or UV	None
Accelerated Weathering:		
After 250 hrs. UV exposure	2 max. (No cracking)	0 (No cracking)
After 250 hrs. UV exposure plus bending 180° at -21°C	2 max. (No cracking)	0 (No cracking)
Adhesion and Cohesion Under Cyclic Movement:		
On Mortar	9cm^2 max bond loss	<3cm^2 bond loss (coh)
On Aluminum	9cm^2 max bond loss	<3cm^2 bond loss (coh)
On Glass	9cm^2 max bond loss	<3cm^2 bond loss (coh)
Peel Adhesion, kn/m:		
On Mortar	0.875 kn/m min.	3.2 kn/m coh.
On Aluminum	0.875 kn/m min.	6.7 kn/m coh.
On Glass	0.875 kn/m min.	6.0 kn/m coh.
On Glass After UV	0.875 kn/m min.	8.0 kn/m coh.

Durability samples made on glass, anodized aluminum and mortar with this sealant were immersed in water at ambient lab conditions for over a year. After the water immersion these samples were tested for cyclic movement at ±25% movement. Although the sealant had softened considerably during the water soak, it passed the cyclic testing with no loss of adhesion.

Despite the fact that this sealant has performed as well as it has, its use is limited because it is based on manganese dioxide. Manganese dioxide is dark in color so sealants based on it are only available in dark colors. Also, when these sealants are exposed to weather, they may cause staining on vertical masonry surfaces.

The following formulation is for a one-part polysulfide sealant based on a new curing agent, sodium perborate. It also incorporates a combination of adhesion promoters to give it good adhesion to glass, anodized aluminum and mortar without the use of primers.

Suggested Starting Formulation For A One Part
Sodium Perborate Cured Polysulfide Sealant

TABLE 4

INGREDIENT	PARTS BY WEIGHT
Polysulfide polymer	100.0
Chain-stopper	0.1
Thixotrope	3.0
Mol Sieve	3.0
Pigment	20.0
Calcium carbonate	30.0
Chlorinated plasticizer	30.0
Treated calcium carbonate	60.0
Phthalate plasticizer	20.0
ELP-612 (Epoxy adhesion promoter)	5.0
ZL-1866 (Acrylic adhesion promoter)	2.5
Silane adhesion promoter	1.0
BaO Paste	5.0
Sodium Perborate Paste	10.0

ELP-612 and ZL-1866 are both modified polysulfide polymers that act as adhesion promoters. Being based on the polysulfide polymer ensures that they will have excellent compatibility with the polysulfide polymer.

ELP-612 is an epoxy terminated polysulfide polymer and ZL-1866 is an acrylate terminated polysulfide polymer. Both of these polymers improve the adhesion to porous and non-porous substrates. The silane is the typical adhesion promoter for improving adhesion to glass.

Sealants based on this formulation have been tested against ASTM C-920-87 as class 25 building sealants on all three substrates and have easily met all the requirements without the use of primers.

TABLE 5

Sealant: One-part $NaBO_3$ cured		**Reference:** DL-9415
Type: S	**Grade:** NS	**Cure:** 1 wk at 25°C/50% RH, plus 1 wk at 38°C/95% RH, plus 1 wk at 25°C/50% RH
Class: 25	**Use:** NT, M, G, A	**Primer:** None

	SPEC LIMITS	TEST RESULTS
Rheological Properties:		
Horizontal at: 50°C	No deformation	No deformation
5°C	No deformation	No deformation
Vertical at: 50°C	4.75mm max. sag	No sag
5°C	4.75mm max. sag	No sag
Extrusion Rate	45 sec. max.	3 seconds
Hardness, Shore A	15 to 50	30
Effect of Heat Aging	10% max. wt loss No cracking No chalking	2.3% wt. loss No cracking No chalking
Tack Free Time	72 hrs. max.	>8<16 hrs.
Stain and Color Change (on mortar)	None, air or UV	None
Accelerated Weathering:		
After 250 hrs. UV exposure	2 max. (no cracking)	0 (no cracking)
After 250 hrs. UV exposure plus bending 180° at -21°C	2 max. (no cracking)	0 (no cracking)
Adhesion and Cohesion under Cyclic Movement:		
On Mortar	9 cm^2 max bond loss	0 bond loss
On Aluminum	9 cm^2 max bond loss	0 bond loss
On Glass	9 cm^2 max bond loss	0 bond loss
Peel Adhesion, kn/m:		
On Mortar	0.875 kn/m min.	2.8 kn/m coh
On Aluminum	0.875 kn/m min.	2.3 kn/m coh
On Glass	0.875 kn/m min.	2.6 kn/m coh
On Glass After UV	0.875 kn/m min.	2.1 kn/m coh

Today, a number of companies are working on developing products that may act as potential adhesion promoters. As these products become available sealant manufacturers are incorporating them into their formulations, and recommending that primers not be needed with their sealants except where the sealants will be subjected to extremely severe conditions.

Definitions of abbreviations:

MnO_2	Manganese dioxide
$NaBO_3$	Sodium perborate
Type M	Multicomponent
Type S	Single component
Grade NS	Non-sag
Class 25	± 25% movement
Use NT, M, G, A	Non-traffic, mortar, glass, aluminum

Glazing

Michael D. Wilson

IN-SITU RE-GLAZING OF STRUCTURAL SEALANT GLAZING JOINTS A CASE HISTORY

REFERENCE: Wilson, M. D., **"In-Situ Re-Glazing of Structural Sealant Glazing Joints-A Case History,"** Science and Technology of Building Seals, Sealants, Glazing, and Waterproofing: Fourth Volume, ASTM 1243, David H. Nicastro, Ed., American Society for Testing and Materials, Philadelphia, 1995.

ABSTRACT: Structural sealant glazing has attracted considerable attention due to its inherent glass retaining method which allows smooth exterior glass walls interrupted only by narrow weatherseal joints. The technique has become increasingly popular with owners and architects since its inception in the early 1970's.

In 1990 it was determined that a five story office building nearing completion had structurally glazed insulating glass units in multi-story bay windows that had been incorrectly installed. Verification of the required structural glazing design was performed to compare with the as-built conditions, and re-glazing the structural silicone glazing joints was indicated to be warranted.

The construction team desired to perform the re-glazing without removing the installed units for cost and time reasons. In-situ re-glazing of a single insulating glass unit was performed as a test of the proposed remedial technique. The test was deemed successful and the remaining defectively glazed units were re-glazed in-situ.

The design and construction circumstances that led to the situation requiring the re-glazing, the lessons learned during their resolution, the in-situ re-glazing techniques utilized, and comments on the need for potential re-glazing of structurally glazed systems are presented.

KEY WORDS: structural sealant glazing, structural silicone sealant, spacer tape, adhesive, architectural drawings, shop drawings, insulating glass units, bay window, curtain walls, glass, aluminum mullions, in-situ re-glazing, weatherseal butt joints

BACKGROUND

The building of concern was a five story office structure located on a resort island outside the United States. The building was constructed on a site along the shore of the island's main harbor. The design wind speed loading for the building exterior was 140 mph [225.3 km/hr]. The shore was exposed to several hurricanes during the 1980's.

The structure consisted of steel and concrete framing, with concrete masonry in-fill exterior walls between floors and between framing members. The concrete masonry walls were covered with the base coat, embedded reinforcing mesh, and finish coat of an exterior insulation finish system.

Multi-floor bay windows containing extruded aluminum vertical and horizontal mullions were attached to the building's concrete and steel frame (see Figure 1). The bay windows extended over either two or three floors. The architect's design required insulating glass units (IGUs) to be held in place by four-sided structural sealant glazing (SSG) to form the smooth glass exterior curtain wall of the bay windows.

Exterior weatherseal joints between IGUs, IGUs and window perimeter framing, and elsewhere in the building exterior were required to be sealed with silicone sealants.

The author was retained by the architect to visit the building as it neared substantial completion, observe the various types of completed sealant joints in the building exterior, comment on the apparent workmanship and compliance with the project requirements, and recommend corrective work if appropriate.

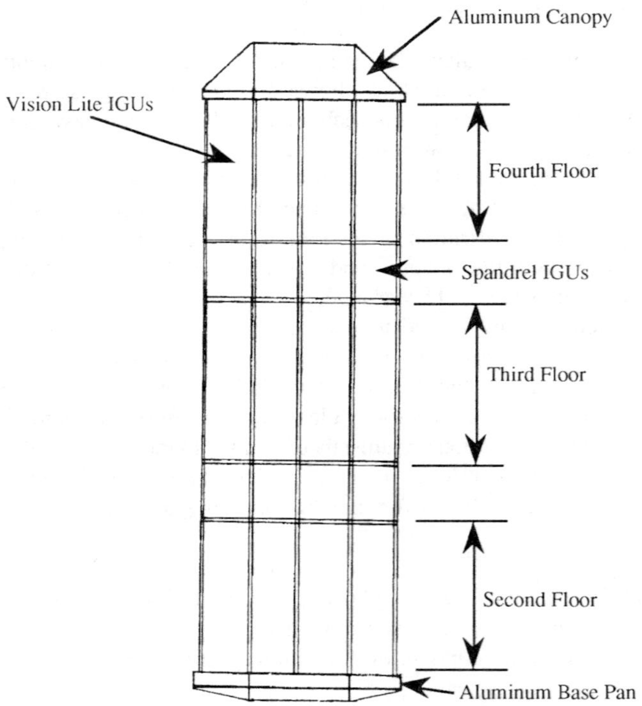

Figure 1: Typical Elevation of Bay Windows

The exterior surfaces of the bay windows were observed via a platform box suspended from a crane. During visual observations of the exterior weatherseal and perimeter joints of the bay windows, the SSG joints retaining the IGUs in place happened to be observable through the glass. The SSG joints appeared to be composed of two materials, an outer (closest to the edges of the IGUs) gray-colored material covered by an inner black silicone sealant (see Figure 2). The apparent thinness of the black silicone sealant (1/16 to 1/8 inch) [1.6-3.2mm] prompted the author's curiosity and a suggestion was made to the architect that the SSG joints be evaluated from the building interior for compliance with the project design and specifications.

The structural glazing joints were observed from the building interior, and test cuts were made in selected locations to evaluate the in-place sealant joint dimensions, construction, and adhesion. The architectural design and window system fabricator shop drawings were reviewed to compare the required installation details with the as-built conditions. The project specifications and related correspondence were also reviewed.

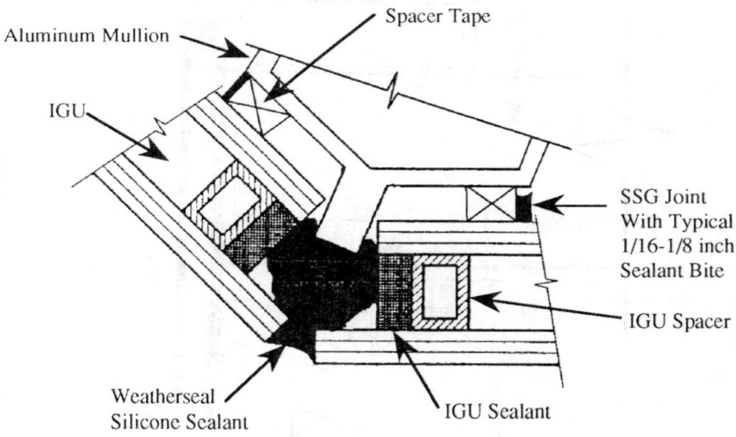

Figure 2: Typical As-Built SSG Joints Before Re-Glazing

The architectural drawings contained one detail sheet presenting the intended design of the bay windows' curtain wall system. Although the appearance of the bay window elevations and details obviously did not use exterior mechanical glazing retainers, terms such as structural sealant glazing joints, structural glazing, sealant bite, glue-line thickness were not present on the architectural drawings. The project specifications stated that "bay windows shall be the butt-glazed wall system, and ground floor windows the surface cover bead system...", providing the only specification reference distinguishing between the intended structural glazing system of the bay windows and the mechanical glazing system of other building windows. The architectural drawings also included one detail showing a concrete floor slab immediately behind approximately the top six inches of the spandrel IGUs. Less than one inch appeared to be present between the edge of the floor slab and the inside lite of the spandrel IGU. Due to the projecting fin on the mullion, it appeared this head joint had to have been structurally glazed through the gap between floor slab and IGU (see Figure No. 3).

The window fabricator's shop drawings approved by the architect included full-scale details of the horizontal and vertical mullions for the bay windows. Although not dimensioned in any of the details, it appeared the shop drawings required a SSG joint bite of 3/4 inch [19.1mm] with a glue-line thickness of 1/4 inch [6.4mm] (See Figure 4). Neither the architectural or the shop drawings indicated the use of spacers, setting blocks, or shims in the glazing details.

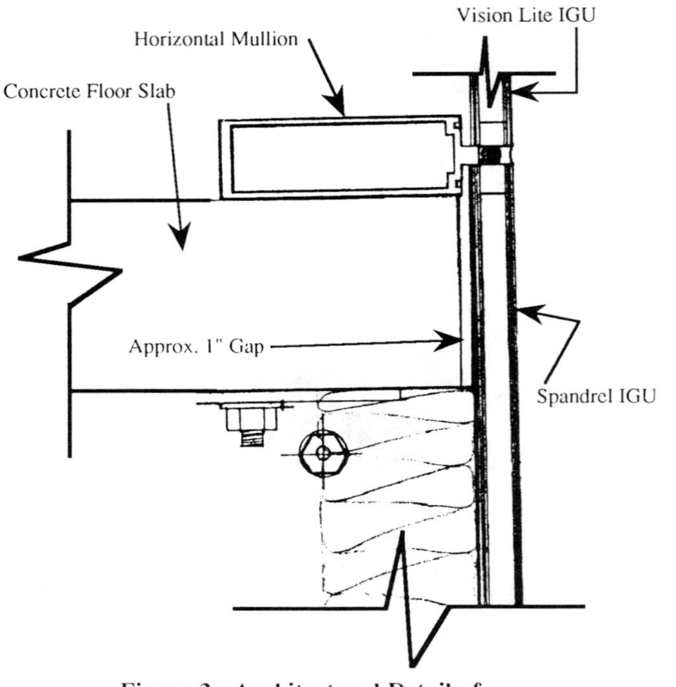

Figure 3: Architectural Detail of Spandrel IGU Head Joint

Test cuts made at the interior perimeters of IGUs throughout the building indicated SSG joints with a bite dimension of 1/16 to 1/8 inch [1.6-3.2 mm] had typically been provided. Observations also indicated that numerous locations existed where placement of the structural sealant in the joint receptacle was incomplete, and in some cases the sealant physically touched only one side (either the glass or the aluminum) of the joint. The sealant appeared to have been placed with the gunned bead being "pulled" along, rather than being pushed so that the sealant would fill the joint. A soft gray foam material was visible beneath the black silicone sealant in these locations. These conditions were typical for almost all of the vision lite IGUs in the bay windows of the building. The spandrel IGUs between floors were observed to contain no silicone sealant for approximately the top eight inches of their vertical sides and all of their top horizontal sides.

Weatherseal joints between the IGUs were shown in the architectural and shop drawings in a typical industry form, indicating the use of backer rods and a weatherseal sealant adhered generally to the outer pane of the IGUs.

Test cuts made into the bay windows' weatherseal joints from the building exterior indicated that the backer rods had been omitted. The joints between opposing edges of IGUs and between the outer edges of IGUs and the exposed perimeter mullion had been almost completely filled with silicone sealant, adhering the IGUs to each other for their full thickness.

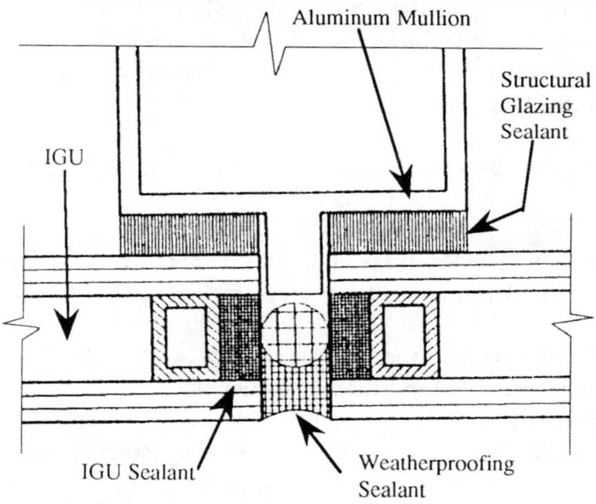

Figure 4: Typical Shop Drawing Detail for Vertical and Horizontal Mullions

The installing mechanic of the glazing subcontractor was questioned regarding the as-built glazing technique versus the details shown in the shop drawings. The mechanic stated that he had created a "combination weather seal and structural seal between" the IGUs, and that the silicone "inner seals were now only cosmetic seals". The mechanic also stated that two-sided adhesive spacer tape (the gray foam material) had been placed in the "inner seals", and was providing sufficient retaining adhesion to keep the glass on the building. The mechanic further related that the two-sided tape had been positioned during IGU placement so that a receptacle yielding a bite of 3/16 inch [4.8mm] or less was provided for the silicone sealant. This was related to be a normal procedure utilized by the glazing subcontractor for windows made by the particular fabricator, and to have been used by the subcontractor in many buildings.

The window system fabricator was contacted to determine what the dimensions of the SSG joints were required to be. A representative of the fabricator agreed that the shop drawings did show a bite of 3/4 inch [19.1mm], but that 3/8 inch [9.5mm] of the adhesion bite was to be provided by a piece of the two-sided adhesive tape. The other half of the 3/4 inch [19.1mm] bite (3/8 inch) [9.5mm] was to be provided by the silicone sealant.

The manufacturer of the two-sided adhesive tape was contacted regarding the structural adhesive contribution of the tape. The manufacturer stated that the tape was to be used only for the purpose of providing a spacer, and should not be relied upon to provide any portion of the structural glazing adhesion of the window system. The tape was marketed and sold as a fabrication aid during glazing only.

During evaluation of the SSG joints, the author calculated the required SSG joint bite for the different sizes of IGU present in the bay windows. Emswiler's velocity pressure equation $Q=0.00256V^2$ (where Q equals the static wind pressure in psf and V equals the wind speed in mph) yielded a wind design load pressure of 50 psf [244 kg/m^2] for the design wind speed of 140 mph [225 km/hr]. The architect provided confirmation that the required design wind load was 50 psf [244 kg/m^2]. The trapezoidal method for SSG joint design was used to determine the required bite:

$$\text{BITE} = \frac{\text{Shortest Unit Dimension x Design Wind Load}}{\text{24 x Sealant Design Strength}}$$

[IGU dimension in feet, wind load in pounds per square foot, and sealant design strength in pounds per square inch. The manufacturer's recommended sealant design stress was 20 psi [14062 kg/m^2].]

The required bite for the IGUs in the bay windows were thus 3/8 inch [9.5mm] for the largest IGUs and 5/16 inch [7.9mm] for the smaller sizes.

The conclusion that the majority of the bay window IGUs were inadequately retained in place was provided to the architect, along with a recommendation that each window be test cut in the SSG joint on all four sides to determine the as-built condition. Reglazing IGUs having less than the required sealant bite was recommended.

PROBLEM DEVELOPMENT

Several interesting circumstances occurred which allowed the inadequate SSG joints situation, discussed above, to develop.

The lead architect in the design of the building left the company to start his own firm prior to the start of construction on the building, and was thereafter not involved with the project. The architect assigned to continue the project (the author's client) related to the author that he had "never heard of structural glazing", and had relied on the window fabricator and installing subcontractor for proper design and installation of the windows.

As previously mentioned, neither the architectural drawings nor the bay window fabricator's shop drawings provided dimensions on the SSG joint details. However, several of the shop drawing details were shown at full scale (see Figures 4 and 5). A review of the project correspondence indicated the manufacturer of the specified silicone sealant had been asked to review the shop drawings, apparently to judge the acceptability of the intended usage

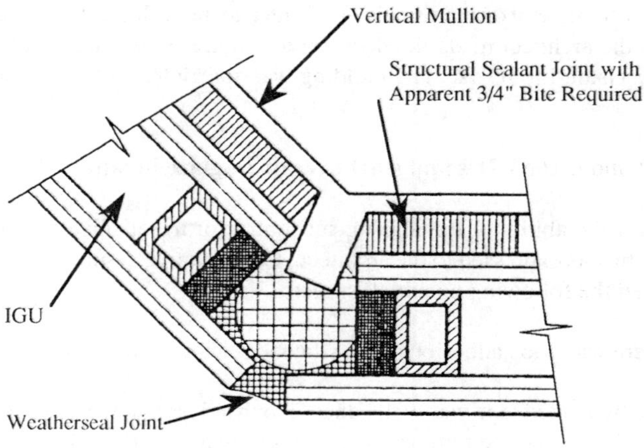

Figure 5: Shop Drawing Bay Window 135° Corner Detail

of the sealant. The sealant manufacturer's technical reviewer provided a letter to the window fabricator (a copy was provided to the architect) stating that "based on a 4'0" [1.2m] x 8'0" [2.4m] maximum lite size and a 20 psi [14062 kg/m^2] sealant design strength the 3/4" x 1/4" [19.1mm x 6.3mm] bite is sufficient to withstand a 60 psf [292.9 kg/m^2] maximum wind load... As per our discussion, the temporary clamping device (i.e. pressure plate) is required to hold the glass into place before the sealant is fully cured. The [double-sided adhesive] tape will be used as a spacer on the mullion and transome."

Apparently the sealant manufacturer's representive determined the design's required SSG joint bite by scaling off the dimension on the full scale shop drawing details. However, the representative did not point out that the total available bite dimension in the mullion design was approximately 13/16 inch [20.6mm], and that a spacer tape of maximum bite 3/8 inch [9.5mm] would have to be positioned at the extreme outer edge of the IGUs to provide a 3/4 inch [19.1mm] remaining available bite for the silicone sealant. The manufacturer's letter also did not state if 3/4 inch [19.1mm] was the minimum bite required or if there was tolerance for less bite.

The glazing subcontractor's installing mechanic reportedly possessed copies of the shop drawings during installation of the IGUs. Whether the window system fabricator provided a copy of the sealant manufacturer's letter to the glazing subcontractor is unknown. Regardless, the glazing subcontractor did not follow the SSG or weatherseal joint details shown on the shop drawings, and quality control inspections by the general contractor or architect were not performed during installation.

PROBLEM RESOLUTION

Two additional consultants and the sealant manufacturer's representative were requested to visit the site and comment on the situation presented to the architect by the author. All three concurred, and the architect made the decision to require re-glazing of all defective SSG joints of all the vision lite IGUs. The building was scheduled to be occupied within five weeks.

Re-glazing: Remove the IGUs and start over or re-glaze in-situ?

Concern over the ability of the glazing subcontractor to satisfactorily re-construct the SSG joints in-situ was expressed to the architect. Conservative assumptions prompting this concern included the following potential issues:

- That there was good adhesion of the silicone sealant to the glass and metal surfaces.
- That the two-sided tape was well adhered to the glass and metal surfaces of the SSG joints and could not be removed without removing the IGUs first.
- That, upon removal of the tape and silicone sealant, the mullion facing onto the joint cavity would not be observable for subsequent cleaning and priming.
- That removal of cleaning solvent residue could not be ensured and that the mullion facing could not be primed without undue risk of primer being applied also to the glass.

An IGU was selected for an in-situ re-glazing test. The glazing subcontractor fabricated a "rake" tool with a cutting tooth to reach into the SSG joint cavity (see Figure No. 6). The rake was used to remove the existing glazing materials and establish a relatively clean 3/4 inch [19.1mm] deep by 1/4 inch [6.4mm] wide joint cavity. The cavity was then cleaned using solvent rag and clean dry rag wiping procedures. An open cell backer rod was placed into the joint to provide a joint cavity of 1/2 inch [12.7mm] minimum bite. The mullion facing was primed, and new silicone sealant was gunned into the joint and tooled. Factors which allowed this action to proceed successfully included:

- The existing silicone sealant in the joints was marginally adhered to the glass and metal surfaces of the SSG joints. It appeared the glass and/or the metal surfaces were dusty when the sealant was installed and the resulting adhesion was nominal. The fact that there was also little silicone sealant present and that it often contacted only one side of the joints aided in the ease of removal.
- The two-sided tape generally did not adhere well to either the glass or the metal face of the mullion, and was readily removed. The apparent dusty condition of glass and metal prior to initial glazing may have contributed to this.

- Sufficient light was available to place a thin flat mirror (3 by 8 inches) [76.2 by 203.2mm] into the joint cavity to see if the existing sealant and tape had been completely removed. The mirror could also be used to view the mullion face after solvent cleaning procedures.

- A thin strip of sheet aluminum or plastic could be placed into the joint against the glass, and primer could then be applied to the mullion face without placing primer onto the glass (and subsequently inhibiting new sealant adhesion to glass).

- The vision lite IGUs were accessible on all four sides with relative ease due to their size and the design of the office spaces.

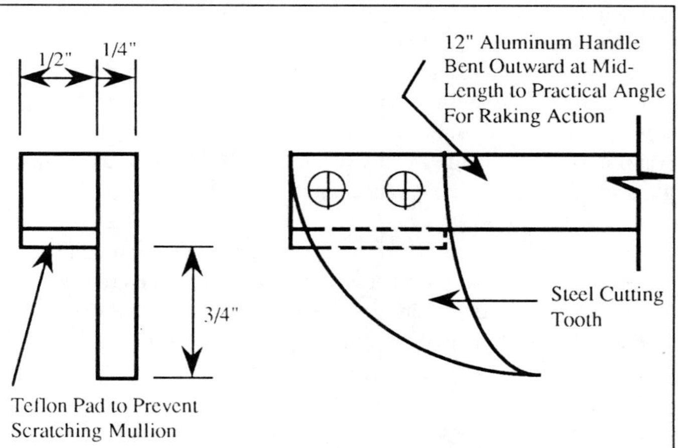

Figure 6: Details of Head of Sealant Joint "Rake"

Execution and Quality Control Testing

Each side of each vision lite IGU was inspected and test cut in at least two locations to assess as-built workmanship and existing SSG joint bite. Joints which were incompletely filled with sealant and/or did not have the required SSG bite dimension were cleaned out and re-glazed. Joints which appeared to meet the required bite dimensions but appeared to have poor adhesion of the existing sealant during the test-cutting were also re-glazed. Approximately 80 percent of the vision lite SSG joints were re-glazed.

The new silicone sealant in the joints was tested for adhesion after passage of the manufacturer's recommended minimum cure time. Utilizing the manufacturer's recommended field testing procedure, all of the quality control tests indicated excellent adhesion was obtained in the re-glazing.

A decision was reportedly made by the building owner to not require re-glazing of the smaller spandrel IGUs in order to avoid destroying completed ceiling areas for access to the spandrels. Possibly due to the mass of sealant in the weatherseal joints adhering the IGUs together, and the successful re-glazing of surrounding vision lites, the owner reportedly accepted an extended guarantee from the general contractor against potential problems with the spandrel units.

CONCLUSIONS AND LESSONS LEARNED

This case history indicates that, if certain circumstances are present, reconstructing structural sealant glazing joints without removing glass panes from a curtain wall is possible. Although the circumstances allowing this result with the vision lites on the subject building may have been luck, thoughtful design can foster the likelihood of such success on other buildings.

The concept of reconstructing structural sealant glazing joints in-situ may prove to be very valuable in the long term. Structural sealant glazing is still a rather young architectural technique. The oldest structurally glazed buildings are now 25 years or more in age, and the silicone sealants which have made the technique possible are not much older. Whether these sealants will retain the strength of their adhesion to glass and metal mullions for the typical 50 year design life of modern buildings remains to be seen.

Several points can be made as "lessons learned" for future projects involving structural sealant glazing. Such points to consider include:

- Although architectural firms may not make the calculations to determine the dimensions of framing members and glazing materials required to meet their design wind loadings, the firm's architect responsible for the building exterior should be well aware of the various glazing methods and techniques in the industry (i.e., how are exterior cladding components such as glass or stone veneer to be secured to the building frame?). This consideration may have avoided the design detail which made it impossible to field structurally glaze the upper eight inches and the entire top horizontal SSG joints of the spandrel IGUs.

- Curtain wall fabricators preparing shop drawings for structural glazing projects should clearly show the placement of all spacers, and clearly indicate all the required dimensions of the SSG joints. This will make it less likely that reviewing manufacturers, general contractors, or architects will misinterpret - or worse, overinterpret - the intended construction. This will also help installing subcontractors to understand clearly what is required of them, the final link in the construction chain.

- General contractors should develop and execute quality control inspections of installing structural sealant glazing subcontractors, as they do with other critical areas in a building's construction. As a quality assurance measure, architects should insist upon submittals from the general contractor indicating the performance of the contractor's quality control effort.

As with the installation of any building exterior wall materials, adequate and thoughtful design, quality materials, and knowledgeable workmanship are required for a properly functioning and safe structurally glazed glass curtain wall.

References

[1] "Minimum Design Loads for Buildings and Other Structures," American National Standards Institute, ANSI A58.1-1982, pp.12-16.

[2] Klosowski, J.M., "Sealants in Construction," Marcell Dekker Inc., 1989, pp. 226-229.

[3] "Silicone Structural Glazing Guide," General Electric Company, 1987, p. 11.

George B. Lowe[1]

EUROPEAN INSULATED GLASS UNIT TESTING

REFERENCE: Lowe, G. B., **"European Insulated Glass Unit Testing,"** Science and Technology of Building Seals, Sealants, Glazing, and Waterproofing: Fourth Volume, ASTM STP 1243, David H. Nicastro, Ed., American Society for Testing and Materials, Philadelphia, 1995.

ABSTRACT: The European Community is pledged to reach a situation where there is no technical barrier to trading between the community partners. To this end the Commité European de Normalisation (CEN) committee TC129 - Glass in Buildings was set up. The Working Group WG4 has the specific task of writing a specification for Insulating Glass Units.

The instructions to the committee were that any insulating glass units currently meeting there own national standard should not be disadvantaged, at least in the first instance, by not being able to meet the new specification.

Naturally National views were put forward in the early stages, and there has been a period of intense negotiation, and compromise, before the specification started to take shape.

The current situation is that although not complete, the specification is now taking shape and a series of Round Robin comparisons have been carried out to verify the tests included.

KEYWORDS: European Community, Insulating Glass Units, Specification, Standards

INTRODUCTION

The formation of the European Economic Community (E.E.C.) has brought together the nations of Europe to form a consumer market similar in population size to that of North America. The Insulated Glass manufacturers of Europe produce in excess of 80 million sq. metres of units per annum. Unfortunately the technical requirements for these units are covered by 10 different specification/quality requirements. Most notable among these are the BSI, [1], DIN, [2] and AFNOR [3] specifications.

At the outset it was decided that the elements of construction were either necessary or mostly aesthetic. The factors that were deemed most necessary were called Essential Requirements.

The Commission for the Community decided that where products were required to meet one or more of the Essential Requirements then that product should be covered by a CEN Technical Standard.

[1] Manager - Technical Marketing and I.G. Specialist, Morton International, University of Warwick Science Park, Sir William Lyons Road, Coventry CV4 7EZ, England

The Essential Requirements relevant to insulating glass are

Safety
Environmental Control
Energy Conservation

Because Insulating Glass Units fell within scope of the Essential Requirements a committee under the control of Technical Committee(TC)129 - glass in Buildings - was set up to write a specification. This committee was designated TC129/WG4 - Insulated Glass.

The task facing the CEN committee is to produce a specification that is acceptable to all parties and does not disadvantage any particular type. The problems facing the committee was that within the community insulating glass units meeting national standards included not only polysulfide based single and dual seal units (the most common system) but also polyurethane, silicone, hot melt butyl and desiccant-containing sealant all in single and dual seal formats. To endeavour to embrace all the above systems with one type test and to ensure that quality was not compromised was a major challenge.

THE NEW STANDARD

Although not yet complete the new standard is rapidly taking shape. The format will be a six (or seven) part specification which will clearly define normative (must be met) and informative (desirable but not mandatory to be met) portions.

The basic format will be:

Part		
	(	Levels of attestation of Conformity
	1	Generalities and Dimensional Tolerances
	2	Initial type test on Air Filled I G Units
	3	Initial type test on Gas Filled I G Units
	4	Methods of Test for the attributes of edge seals
	5	Audits and initial inspection of the factory and factory production control
	6	Factory production control.

The various parts will be described together with their current positions.

Part 0 Attestation of Conformity

There is potential for a number of levels of attestation varying from complete self-certification and control to third party certification and audit control.

The CEN committee responsible for the I G specification (TC129/WG4) has recommended that the highest level - full product certification by third party attestation - should be adopted. This is still however under debate and no work is possible on this part.

Part 1 Generalities and Dimensional Tolerances

This section of the specification should have been relatively straight forward. The main points were to be the description of insulated glass unit types covered by the specification and the size tolerances.

There is however a controversial section that relates to edge deflection.

Because of the different ways in which edge deflection had been calculated, it was discovered that there was apparently a wide range of allowable edge deflections ranging from L/175 in the U.K. to L/300 in Germany. Similarly in the U.K. there was no maximum limit specified whilst in Germany on 8mm maximum was included.

The final outcome was to put frame deflection into an informative annex and have a compromise deflection level. The actual wording is:-

"Relevant frames within terms of this standard.

Within this standard a relevant frame is understood as one:

- where the four sides of the insulating glass unit are covered by fixing beads or any other part of the framework,

- and where entrapment of stationary water, long term condensation and/or enhanced water vapour pressure against the seal of the insulating glass is prevented in order to avoid chemical and/or physical attack on the components of the insulating glass and to prohibit excessive water vapour penetration into the insulating glass,

- and showing a sufficient stiffness to limit the edge deflection of the insulating glass under loads such as wind and snow to avoid reduction of life time due to such as loss of adhesion of the insulating glass components. A means of obtaining a recommended minimum frame stiffness is when the frame, considered over the length L of the glazing, together with an infill of low edge rigidity, such as a single glass, shows a relative deflection of L/200 or less with a maximum of 8mm when exposed to design loads, either as single load or as combined load. The design load shall be according to:

  - ENV 1991-1. Eurocode 1: Basis of design and actions on structures. Part 1 Basis of design. Milan, 28th of May 1993.

  - CEN.=/TC 250/SC 1/1993/N 106

  - ENV 1991-2-1 Eurocode 1: Basis of design and actions on structures. Part 2.1 Densities, Self-Weight, Imposed Loads, Snow Loads and Wind Actions and where $\delta_F = 1$ as value for the partial factor for the load."

Part 2 Initial Type-Test on Air-filled Insulating Glass Units

On the surface this part of the specification is relatively straight forward. The units of the defined type are subjected to 56 cycles between 53°C ± 1°C and -18°C ± 2°C followed by 1176 ± 4 hours (7 weeks) at 58± 0.5°C at greater than 95°C RH. Following the test the units must meet certain water absorption criteria.

The simplicity of the statement however belies two years of intense work to establish accurate control conditions and definitions to enable the test to be reproduced in any European test house. To date there has been two European and two UK Round Robin series of tests to try and satisfy users that the test is reproducible. A number of points should be clarified about the test in order to explain why it took so long to develop.

Firstly, the severity of the test substantially eliminates single seal units, for those manufacturers making single seal a lead in time has been negotiated to allow them time to change to dual seal. Secondly the test units must be made to a high standard if the producer wishes to meet the requirements.

Although this part will shortly be on circulation for public comment there are still some problems to be resolved. The area of main concern is the organic spacer incorporating desiccant. Although the Karle Fischer method is outlined, the Round Robin tests have not generated sufficient agreement for the method to be used with any confidence. A further problem is that only small samples of material (0.5 gms) are taken, and this therefore assumes that the water is evenly distributed throughout the system. There is only little evidence to show what the variation in water content as used in factories is likely to be, and the test to demonstrate fit for use is cumbersome. However after four years of trying no improved method has been found.

Regardless of the methods of manufacture, it is intended that periodically cross checks will be made between test houses to ensure reproducibility of results is maintained.

Part 3 Initial Type-Testing on Gas Filled Insulating Glass Units Gas Leakage Rate

The first thing to remember regarding this part is that it is essential to meet the requirements of Part 2 before attempting to meet Part 3.

The main assumptions of Part 3 are that the majority of Gas filled units will contain Argon, Krypton or Sulfur hexafluoride or a mixture of these together with residual air. Any other Gas will require more detailed testing to show that there is no reaction between it and the edge seal components.

The main requirements are that the level of Gas fill will be measured and shall lie between +10% and -5% of that stipulated.

The gas loss rate is defined as being the proportion in parts per thousand by volume of gas leaking from a gas filled unit per year (10^{-3} a^{-1})

Again strict conditions of measurement are applied and in order to obtain meaningful results the unit manufacturer will be required to record the temperature and pressure (atmospheric) at which the units were sealed.

Round Robin testing for gas fill and gas loss rates have shown reasonable reproducibility. The only point of contention is the manner by which the leaked gas is accumulated prior to analysis.

Some workers prefer to seal the system and allow the gas to accumulate in the unit cassette, whilst others believe it is more accurate to collect the gas in the liquid nitrogen cooled trap to prevent gas concentrations around the unit influencing the result. This point is so far unresolved.

Part 4 Methods of Test for the (Physical) Attributes of Edge Seals

Parts (3 constitute what is commonly referred to as standard insulating Glass Unit testing. Parts 4 - 6 are a departure in that there is now included with the specification programme tests for the main constituents parts to ensure repeatability and conformity.

The normative section deals primarily with adhesions. In order to cover all types of edges seal, the test is only concerned with the stresses and strains imposed on the I.G. unit edge during testing. It is anticipated that no failure (adhesive or cohesive) will be allowed within the area of test.

Even such simple testing has created problems in that the softer hot melt systems have difficulty reaching 50% extension without some evidence of cohesive or adhesive failure.

The informative tests are listed to standardise the manner by which data is presented by the sealant manufacturer to the unit manufacturer. The difficulty in drafting these proposals were to avoid the interpretation that certain parameters (e.g., high modulus) were of necessity an improvement when compared to sealants having a lower test value.

The test programme is however deemed useful for comparing sealants of the same generic type.

It has to be remembered that the tests required by Part 4 are over and above any information supplied in the form of a certificate of conformity.

It is anticipated that some form of fogging test will be incorporated within Part 4 as a normative section, as it is felt that without such a test the whole specification will be devalued.

Part 5 Audits and Initial Inspection of the Factory and Factory Production Control

Little work has been done on this section, but it could have an important bearing on the way manufacturers operate depending on the time between type tests.

If the audit is third party assessed, as required by the committee, a series of tests will be included to ensure continuing compliance. Suggestions have been put forward regarding the use of Thermo Gravimetric Analysis, Infra-Red Spectrophotometry etc. as well as adhesion, mass loss and creep tests on units.

Part 5 will only be completed when Part 0 is finished and the level of attestation is known,

Part 6 Factory Production Control

This part of the specification is mandatory and the Construction Products Directive (CPD) provides that the manufacturers may use the EC conformity mark for their construction products only if they have a production control system to ensure that the product conforms to the relevant technical specification (Article 13.3A)

As can well be imagined this part of the specification has created a number of problems, not least of all is laying down a minimum level of operation to be followed by all I.G. unit manufacturers regardless of size.

The main outline of Part 6 is written along the lines of EN 29000 (ISO 9000) and anyone familiar with these will recognise the general format regarding management, quality systems, documentation, training etc.

The controversial aspect has been to specify some of the testing and its frequency as relates to I.G. Unit manufacture.

It has to be remembered that as with Part 4 the testing

outlined is mandatory and is over and above the information on the certifications of conformity normally supplied by the component manufacturers. It also creates pressure on the component manufacturer to operate to ISO 9000 and to supply data in the form required by the specification rather than the way he has traditionally supplied data. The additional information required can be carried out by the component supplier and incorporated in the certificate of conformity.

The committee has however tried hard to make Part 6 operable and intelligent and allows a degree of freedom within ISO 9000 to accommodate the different methods of manufacture. An important aspect that only products that meet the requirements of Part 4 and Part 6 are allowed onto the production line. Failure to observe this fact could invalidate the certificate.

There is an implied element within the specification that for various reasons IG unit manufacturers may want to change one component within their system. It is envisaged that provision will be made to allow change providing that the new product is of the same type and is already used in the proposed combination in an approved system. The flexibility to change has still to be resolved, but it is not the intention of the specification to prevent sensible development of the IG unit by the manufacturer.

CONCLUSION

As can be seen the new specification differs from normal type-test specification in that the production and product control is included in the normative parts. Thus it is no longer sufficient just to test units to a type test and claim compliance.

The specification will start to come into force during 1995 and by 1998 will be complete. Anyone wishing to sell IG units within Europe will have to comply with regulations and meet the tests. Deemed to comply will no longer be allowed.

It is believed the new standard will be of considerable benefit to the consumer as not only will the products be tested uniformly throughout Europe, but also to a more stringent specification than any of the current national standards which will automatically be superseded.

REFERENCES

[1]BS 5713 : 1979 British Standards Institute Specification, "Specification for Hermetically Sealed Flat Double Glazing Units.

[2]DIN 1286 Deutshes Insitute fur Normign e.v. "Insulating Glass Units : Air Filled Quality Specification for Ageing".

[3] P78-451 Association Francaise de Normilisation "Insulated Glazing, Tests of Resistance to Penetration of Moisture".